Custer's Other Brother-in-Law, The Orders Book of Lieutenant Frederic S. Calhoun

By Thomas R. Buecker and Meagan Huff
Edited by Jefferson Davis

Published by Norseman Ventures LLC, Vancouver, WA
Printed by Sun Graphics, Kansas

ISBN: 9781893186316
Library of Congress Number: 2015911744

Copyright 2015 Jefferson Davis, and Meagan Huff.
Except:

Frederic S. Calhoun A little-known member of the "Custer clique', by Tom Buecker © 1994

The Frontier Years, by Colonel Thomas Anderson is an excerpt from a longer article, the 14th Regiment of Infantry, published in *The Army of the United States Historical Sketches of Staff and Line with Portraits of Generals-in-Chief.* Rodenbough, Theophilus F. Bvt Brigadier General U.S.A. and Haskin, William L. Major, First Artillery, editors. Maynard, Merrill, & Co. New York, 1896, within the public domain.

Scanned images of the Frederic Calhoun Orders Book are reproduced with permission of the National Park Service. The rights of other photos, illustrations, etc. reproduced herein are noted within this book.

Buecker, Thomas R. and Huff, Meagan, Edited by Jefferson Davis
Custer's Other Brother-in-Law, The Orders Book of Lieutenant Frederic S. Calhoun
 1. Calhoun, Frederic
 2. U.S. Army, late 19th Century
 3. Custer, George Armstrong
 4. Includes Index

Acknowledgements

This book could not have been published without the assistance and patience of many people. Among them are Mrs. Elizabeth Bonney, who donated the Frederic Calhoun book and several family photographs for preservation and historic interpretation. Ms. Tracy Fortmann, National Park Service Superintendent of the Fort Vancouver National Historic Site, which includes Vancouver Barracks, Washington; who gave permission for scans of the original book pages to be published. Thanks to Theresa Langford, curator at the Fort Vancouver National Historic Site, for preserving the Calhoun family books, and smoothing the way for the publication of this book. Thanks to Meagan Huff, who took on the onerous task of transcribing the original handwritten and printer notes in Frederic Calhoun's book into electronic text. Thank you Zina Krivoruk for proof reading the final draft of this book. Thanks to my nephew Matt D. Parker for his help on the cover for this book.

A very special thanks to Mrs. Kay Buecker, who gave permission to reprint of her late husband Tom Buecker's biography of Frederic S. Calhoun in this volume. Thanks to the many historians and mueologists in the Midwest who helped accomplish this, as well as suggested edits and corrections on the draft copy of this book. Among these people are, Mr. David Bristow and Mr. Jim Potter, of the Nebraska State Historical Society, and many others. Even if I did not mention you all by name, I owe you a great debt.

Jefferson Davis
Editor

Table of Contents

Foreword..1

Editor's Foreword...2

Introduction..4

Lieutenant Frederic S. Calhoun Biography..................9

History of the 14th Infantry Regiment 1865 to ca 1895 27

Timeline of Lieutenant Frederic Calhoun's Career..37

Map of Western United States...................................40

Appendix A: Frederic S. Calhoun's Orders Book....43

Glossary of Acronyms and Abbreviations...............234

Appendix B: Lineage and Honor of the 14th Infantry Regiment..236

Bibliography..242

Index..243

Foreword

The Nebraska State Historical Society (NSHS) and the field of western American history lost a longtime friend and colleague with the death of Tom Buecker on February 2, 2015. Buecker came to work for the NSHS in 1977, serving as curator at the Neligh Mill State Historic Site (1977-1985), the Fort Robinson Museum (1985-2011), and the Thomas P. Kennard House in Lincoln (2011-2015). He authored three books: *Fort Robinson and the American West, 1874-1899* (1999); *Fort Robinson and the American Century, 1900-1948* (2002); and, *A Brave Soldier & Honest Gentleman: Lt. James E. H. Foster in the West, 1873-1881* (2013), which received an Award of Merit from the American Association for State and Local History. He also wrote more than seventy published articles on the history of the West. A fourth book, *Last Days of Red Cloud Agency*, will be published by the NSHS in 2016.

Historians, students, museum visitors, and many others benefited from Tom's generosity with his time and expertise. As he said in a 1985 interview with the *Lincoln Journal Star*, "When a historian uncovers facts about a subject . . . I feel he should share it. Why keep it to yourself?"

David Bristow
Associate Director / Publications
Editor, Nebraska History
Nebraska State Historical Society

Editor's Foreword

I echo the sentiments of David Bristow, Associate Director/ Publications Editor, of *Nebraska History*. While I never met Tom Buecker face to face, I benefited from his openness and generosity. In 2011, when I decided to compile and publish the orders book of Lieutenant Frederic S. Calhoun, I approached Tom via email, asking for a copy of a Fred Calhoun biography, Tom published years earlier in the journal, *Greasy Grass*. Tom sent me one of his few remaining personal copies, and I think he would not have charged me for the copy or postage, had I not reminded him.

We exchanged more emails, where I complained about the small number of original biographical documents on Fred Calhoun. Tom suggested I consult the National Archives. He soon provided me an archive box number, and point of contact to request copies. In short order, I had a complete set of original documents relating to Fred Calhoun and the times in which he lived. Unfortunately, I put the project on hold, as life got in the way of writing and publishing. When I resurrected this book project, kindness and cooperation followed me.

Sadly, Tom Buecker died recently, and could not consult with me on an expanded biography. Fortunately, his colleagues and friends put me in touch with his wife Kay, who generously gave me permission to reprint the original biography. I have faithfully reproduced it in this book. I wish to thank you all for your kind assistance; as well as other friends and sources of aid, which includes the original copy of Fred's book.

In the Autumn of 2008, I was assigned as historian to the 104th Division, an Army Reserve unit stationed at the Vancouver Barracks, in Washington State. In that time, I acted as the de facto historian for the post. One day, a soldier came into my office, he told me, two visitors wanted to see a historian about some books they owned. That is how I met with Kent and Elizabeth Bonney.

They showed me two very old, ledger style books, saying they belonged to Mrs. Bonney's fore bearers. They asked me if I had ever heard of Lieutenant Frederic Calhoun. Fortunately for me, I was doing research for a battlefield tour my unit would take to the Little Big Horn. I said, "Isn't he George Armstrong Custer's other brother-in-law?"

They rewarded me with smiles on my knowledge of military history. She was a niece of Fred Calhoun and his wife; who had one child, a daughter. Their daughter never married; and after her death, family members divided many of the Calhoun's possessions as keepsakes. Mrs. Bonney donated the books to the Army, as long as they stayed at the Vancouver Barracks. A few years later, the US Army left the Vancouver Barracks, and at the Bonney's request, I conveyed the books to the National Park Service, who now own a large portion of the former military reservation. However, in the course of my duties, I had scanned the two books.

One book was Fred Calhoun's, a collection of various orders and special assignments he received while he was in the Army. The other book contained the household accounts of his wife, Mrs. Emma Reed Calhoun. This particular volume includes accounts from the years they were stationed in the Pacific Northwest. When I told the Bonney's I wanted to publish both volumes as research tools for historians, we met and I interviewed Mrs. Bonney about Fred and Emma, Calhoun. They gave me pictures and newspaper clippings, detailing the lives of several Custer Clique members.

I decided to publish Fred's book first because I thought it would be easier. It was not as easy as I thought; while several pages are typewritten, several other entries are hand written by Fred Calhoun and others. In addition to preserving and curating the books, Ms. Meagan Huff, of the National Park Service took on the task of transcribing the sometimes nearly impossible to read notes scribbled by Fred Calhoun. With her help, I have laid out original entries and transcriptions facing each other, so researchers can get a feel for those days.

Jefferson Davis
Major, US Army Reserve
Retired

Introduction

Many Americans know the name of George Armstrong Custer, even if they do not know exactly why. Most historians know the name and the debate surrounding Custer both as a military commander and a human being. Fewer historians know the names of the other US Army soldiers and Native Americans who lived and died at the Battle of the Little Big Horn. Most will have heard of Sitting Bull, and Crazy Horse, of Marcus Reno and Frederick Benteen. Fewer know the names of family members who died with or near George Custer, like his brother Boston, or his brother-in-law James Calhoun; whom some called the Custer Clan, or the Custer Clique. Very few know about George Custer's other brother-in-law, Frederic Sanscay (or Sanxay) Calhoun.

One of the reasons they do not remember Fred Calhoun, is because he did not die at the Little Big Horn with Custer and his extended family. Fred survived the Great Sioux War, while serving as an officer with the 14th Infantry Regiment. Because of his survival, Fred Calhoun faded into obscurity. To better understand Fred Calhoun's story, this book includes a wonderful biography, written by historian Tom Buecker, reproduced with the consent of his wife Kay. It is Fred Calhoun's relative obscurity which we hope makes this book an asset to military history researchers.

Far too many historical researchers rely on secondary sources of information, trusting another historian to blow dust off original documents, and correctly interpret them in essays and books. This can lead to misinterpretations, mistakes and critical errors. Other researchers relying on these erroneous publications can compound those errors in their research. This may be especially true in the case of publications even remotely connected to the Battle of the Little Big Horn.

In the decades since the battle, historians have combed private collections, looking for undiscovered original documents. They have done their work so well, most papers have been discovered. It is a rarity, to find artifacts like the Calhoun family ledger books unpublished. I must point out that this book will not shed much light on the final moments of George Armstrong Custer and James Calhoun at the Little Big Horn; Fred Calhoun was many miles away during the battle. However, his life and career are a backdrop of Army culture and practices during the late 1800s.

In this one volume, we have the administrative history of an infantry officer of the late 19th century. His experiences were mirrored by many other army officers. When seen together, these orders paint an interesting and sometimes unexpected picture of military life. Particularly the drudgery, after all, it cannot all be battle. We hope other researchers will see these same anomalies and use them as starting points for specific research on Fred Calhoun or other military subjects.

For instance, in a career which spanned 15 years, Fred Calhoun participated in 26 courts martial. That is nearly two a year. Was Fred chosen because of some special military or clerical skill? Was there a duty roster, and he served on courts martial when it was his turn? If that is true, what does that say about the number of courts martial convened by the US Army in the 19th century?

For the most part, Fred was circumspect about what he put in his orders book. In two different entries, he wrote the word "personal" in large letters, as if he expected someone to pick up his book and begin reading it. The number of personal comments or asides are very few. This can be frustrating, yet understandable to some historians, who view Fred as a person with feelings.

During the Great Sioux War of 1876, on June 24, [page 25], he and his unit joined the Big Horn & Yellowstone Expedition. The next entry on dates to Sept. 16, 1876 [page 26]. In the time between those two dates, were the deaths of his brother and other Custer Clique members, and the privations Fred Calhoun and his unit endured on the Starvation March. He felt so strongly about that experience, Fred recorded the final speech made by General Crook, when he formally disbanded the expedition on Oct. 24, 1876 [pages 28-30.]

Fred Calhoun is also silent on the death of Crazy Horse on Sep. 5th, 1877. Fred Calhoun was Post Adjutant at Camp Robinson when Crazy Horse was brought into the post . A guard from the 14th

Infantry stabbed Crazy Horse with a bayonet. After Crazy Horse was wounded, they carried him into Fred Calhoun's office, where Crazy Horse died five hours later. There is no note or notation in Fred's orders book between Aug. 8th and Oct. 26, of 1877, just a blank page.

Another interesting facet of this book is the interconnected nature of the Frontier Army of Fred Calhoun's day. Despite the distance between posts and soldiers, if only by paperwork, many individual soldiers touched each other. The index of this book contains the names of over 150 people who at one time or another are mentioned in Fred Calhoun's book.

There are other, less weighty subjects that might inspire historic researchers. For instance, the importance of coffee to the army, even then. One of Fred Calhoun's first missions was to conduct new recruits to their first assignment. He received orders to provide each man with two quarts of fresh coffee a day [page 12.] One board of survey he participated in investigated that quality of coffee beans delivered to outpost garrisons [page 15.] One of the principal concerns of the canteen created by the 14th Infantry Regiment in 1886 was brewing an affordable cup of coffee. The entire report of the Canteen Committee is included within this volume. We hope others are curious enough to try to these and other questions they might ask while surveying this book.

Frederic S. Calhoun
A little-known member of the "Custer clique'
By Tom Buecker
(Originally published in the journal *Greasy Grass* Vol. 10, May 1994)

The key players in the Little Big Horn saga are all familiar to those who study the battle closely: the Custers, Sitting Bull, Myles Keogh, Crazy Horse, Frederick Benteen, Marcus Reno, John Martin and John Ryan. But the battle and overall Sioux War involved thousands of participants, each carrying a story of victory, defeat and personal loss. One of the little known personalities in this great conflict was Frederic Sanscay Calhoun, the brother of Lt. James Calhoun and member of the "Custer clique." Fred Calhoun came close to, but missed, being on that ridge that "mighty afternoon," yet still played an unusual role as the Sioux campaign wound down a year later.

Fred Calhoun was born on April 19, 1847, in Cincinnati, Ohio. His father, a successful merchant, provided the family with more than moderate means. He had one older brother, James (b. Aug. 24, 1845),and a sister Charlotte. The Ohio Calhouns were from "the Calhoun family of South Carolina made famous through Vice President John C. Calhoun."[1] Calhoun began schooling at age 8 and continued until he was 14, when he was sent to Iowa to learn the hardware business. However, an unknown sickness at home compelled him to return to Cincinnati. [2]

In January 1864, brother James enlisted in the 14th Infantry and was soon promoted to company first sergeant. Following in his brother's footsteps, Fred, barely 17 years old, enlisted in the 137th Ohio Infantry, one of a number of "one hundred day" volunteer regiments organized to defend Washington while the Army of the Potomac was sparring with Gen. Robert E. Lee's forces in Virginia. Fred served in Maryland with Company B. His total service in the Civil War equaled that of the 137th, from May 12 until he and the regiment were both mustered out on Aug. 19 "by reason of expiration of service."³ After completing his army service, the younger Calhoun returned home to Cincinnati.

Meanwhile, James continued serving with the 14th Infantry until he was commissioned a second lieutenant in the 32nd Infantry in 1867 at Camp Warner, Oregon Territory. He served in the southwest until assigned to the 7th U. S. Cavalry on Jan. 1, 1871. Fred followed his brother to the west after the war. By his own account, he was "employed at various times as clerk in the Quartermaster Department on the western frontier."⁴

After assignment to the 7th Cavalry, James was stationed with his company in Kentucky. In 1872, he married Margaret Emma Custer, the sister of George and Thomas Custer. In the spring of 1873, the 7th transferred to Dakota Territory and that summer escorted the Northern Pacific Railroad survey crew into the Yellowstone Valley. Fred joined the expedition as an employee of Custer's old West Point classmate and Civil War adversary, Thomas Rosser, chief surveyor for the Northern Pacific. Throughout the long days and weeks in the field, Fred had ample opportunity to establish closer ties with Custer and other 7th Cavalry officers. In fact, he became an unknowing pawn in one of the several difficulties that arose between Custer and Col. David S. Stanley, the expedition commander.

On July 8, Custer allowed Fred Calhoun the use of a horse belonging to Company H. With the company already short-handed of horses, its commander, Capt. Frederick Benteen, objected to no avail. He later complained to Lt. Patrick Ray, expedition

Fred or James Calhoun during the Civil War
Courtesy Elizbeth Bonney

commissary, that Custer was letting a civilian use a government mount while Ray was forced to ride a mule. Stanley, learning of the horse loan, placed Custer in arrest for unauthorized use of a public animal.

Custer countered that Stanley himself had allowed a *New York Times* correspondent to ride an army horse. Also, Custer, who was riding his private mount, reasoned that he had loaned Calhoun his own public mount, which he did not need. Although the incident soon blew over, it only added to Benteen's animosity toward his lieutenant colonel and members of the "Custer Clan."[5]

After the expedition broke up, Custer had both brother Tom and brother-in-law Jim Calhoun put on detached service at Fort Abraham Lincoln, while their company remained at Fort Rice with Benteen. Fred also spent the winter of 1873-74 at headquarters, living with the Calhouns. Of that winter, Elizabeth Custer would later write, "We had such a number of my husband's family in garrison that it required an effort occasionally to prevent our being absorbed in one another."

She added that George's brother Boston came to visit, "and our sister Margaret's husband had a sister and brother at the post." [6]

Early in 1874, Second Lt. John Aspinwall mysteriously disappeared from the 7th Cavalry, and possibly deserted, leaving the regiment with a vacancy. That meant a civilian could be appointed to fill it, not an uncommon act. During the Indian Wars about 12 percent of the army's officers received commissions by appointment as civilians rather than through West Point or from the ranks. By this time, Fred had apparently created a favorable impression among Fort Lincoln's officers and, more than likely, with the 7th's commanding officer, George A. Custer.

He was evidently encouraged to apply for a commission to replace Aspinwall. In early spring, a petition was circulated among the post officers endorsing Calhoun's application. In glowing terms, the petition informed Secretary of War William Belknap that Calhoun was "a young gentleman of superior ability, of excellent habits, and one whose social character is such that added to his fitness for the position renders him a person whom we would gladly welcome to our profession."[7] The document was signed by 12, 7th Cavalry officers, three each from the 6th and 17th Infantry, the post quartermaster, and Lt. Col. Custer.

Not everybody was enthusiastic about the petition. Lt. Charles Varnum approached Benteen, who was visiting Fort Lincoln, to secure his signature. Benteen objected on the grounds that the three-month statute of limitations for Aspinwall's return had not expired. In addition, he asked Varnum, "Don't you think that we have quite a genteel sufficiency of this clique in the Regiment?" Benteen's decision not to sign was soon known in the Custer and Calhoun households. Benteen later stated that several other officers objected and did not sign.[8] Nevertheless, the petition and Fred's application for a commission were forwarded to the secretary of war.

While the bureaucratic machinery moved slowly, Calhoun continued his close association with the "Custer Clan" and in the summer of 1874 accompanied the regiment on its Black Hills expedition. Pvt. Theodore Ewert mistakenly noted in his diary that a brother of Custer (Boston) was appointed forage master, while a brother of Lt. Calhoun was assigned as master of transportation. Actually, Boston was classified as guide, and Michael Smith was master of transportation. Then sarcastically commenting on nepotism, he wrote that the Custers were "thus keeping, as the reader can see, all the paying situations in the family."[9] While in the Hills, Boston wrote to his cousin, Emma Reed, and mentioned Fred Calhoun.

Throughout the summer and fall, Calhoun and his supporters awaited a reply on his application. On Dec. 19, Custer tersely telegraphed Belknap, "Calhoun is here. Can't you get him in Cavalry?"[10] Finally, on Dec. 29, orders were issued from department headquarters for a board of officers to meet at Fort Lincoln to examine Calhoun and any other candidates for appointment as second lieutenants. The board consisted of four officers-and a stacked deck- Lt. Col. William P. Carlin, 17th Infantry; Surgeon James F. Weeds; Capt. George B. Dandy, AAQM; and Capt. George

Yates, 7th Cavalry. Interestingly, three of its members, Carlin, Dandy, and Yates, had also signed the Calhoun petition.11

The board's examination of Calhoun during January 1875 included a thorough medical checkup and a series of written tests on his command of English grammar, arithmetic, geography, history, and the Constitution and government. On the five written tests, which carried a maximum of 44 points, Calhoun scored 26, four points above the minimum acceptable performance. Finding Calhoun in good physical condition, the board found him "a proper person, morally, mentally and physically to be commissioned in the Army."12

On March 10, 1875, Calhoun was appointed a second lieutenant, but as fate would have it, not in the 7th Cavalry but in the infantry. On April 1, he was assigned to his brother's old regiment, the 14th Infantry, stationed in the Department of the Platte. That summer he traveled to department headquarters in Omaha, Neb., for transportation to his first post, Camp Douglas, Utah Territory. On July 5, the new lieutenant boarded a westbound train for Salt Lake City accompanied by 176 new recruits for his regiment. Arriving at Camp Douglas, he was assigned to Company F, commanded by Capt. Thomas Tobey. 13

2nd Lt Frederic Calhoun no date
Courtesy Elizabeth Bonney

Calhoun's duty with his company was short. Owing to a shortage of available officers, he was detached to Company C on Aug. 22. Two days later, that company was ordered to temporary duty at Corinne, Utah, where non-Mormon settlers had become alarmed when large numbers of Shoshoni and other Indians, encouraged by Mormon missionaries, moved into the Bear River Valley north of town. The townspeople feared an outbreak or massacre, but with the troops' arrival, the Indian camps broke up. Company C stayed at Corinne until returning to Douglas on Sept. 23. In February 1876, Calhoun was reassigned to Company F.[14]

In early 1876, the action was far from Utah. In preparing for the summer campaign, Custer gained the transfer of the 7th Cavalry companies in the South to Dakota Territory but on March 30, the adjutant general, for whatever reason, refused his request for assignment of Fred Calhoun to the regiment undoubtedly sparing Calhoun from sharing his brother's fate at the Little Big Horn.[15]

After the Rosebud battle of June 17, 1876, Brig. Gen. George Crook called for more reinforcements and supplies. At Camp Douglas, Companies C, B, F and I of the 14th boarded a train for Medicine Bow, Wyoming Territory, on June 24. From there, the infantrymen marched overland to Fort Fetterman, arriving on the 30th and joining three other infantry companies under the command of Major Alexander Chambers, 4th Infantry. At 7 a.m. July 4, the infantry troops and supply wagons headed north to Crook's base camp near Cloud Peak. So began Fred Calhoun's involvement in the Great Sioux War.

Still unaware of events to the north, the supply and reinforcement column made camp at old Fort Reno on July 8. There a dispatch rider brought news of the Custer disaster, and Calhoun learned of his brother's death. Others shared his grief. Also with the infantry column was Lt. Albert B. Crittenden, 4th Infantry, a cousin of Lt. John Crittenden who died with Custer.

The next day one of the officers recorded, "The command were feeling quite melancholy, and the principal topic of conversation was the loss of Gen'l Custer and command. Lt. Calhoun lost a brother in the fight."[16]

In a later interview with early Custer researcher Walter M. Camp, Valentine McGillycuddy, the surgeon with the infantry column when it left Fort Fetterman, misled and confused historians with a bogus story on when Calhoun and he first heard of the Custer defeat. He claimed that the two men had been sitting at the trader's store at Camp Robinson, several hundred miles east and south, when they learned from Indians there of the great battle. Both men were never at Camp Robinson at the same time until late in October. McGillycuddy's embellishments have been unfortunately accepted by some students of the battle.[17]

Alexander's column continued north, arriving at Crook's camp on July 13. Calhoun received a letter from his brother-in-law, Myles Moylan (who married Charlotte, another example of 7th Cavalry intermarriage), which described what he knew of the battle and its aftermath. Moylan told how the bodies of James and Crittenden, who was on duty with Company L, were both found with their troops.

Moylan also reassured the family: "I was present when Jim was buried and recognized him at a glance," adding that Boston Custer and Autie Reed were both recognized. Moylan concluded, "I would like to write your mother but can't." Calhoun sent the letter to the Reed family in Monroe, Mich.[18]

In August, Calhoun began campaigning with Crook, as the reinforced command cautiously advanced toward the Yellowstone River and an eventual rendezvous with Terry's column. On the 17th, Crook's troops reached the junction of the Powder and Yellowstone rivers. After meeting with Terry, Crook decided to pursue Indian trails east and south. Eventually, with only two day's rations, he turned south, following a fresh trail toward the Black Hills. Thus began an ordeal known as the "Mud," "Horsemeat," or "Starvation" march, which lasted from Sept. 5 to 13. After rations ran out, the troops subsisted on horsemeat.[19]

On Sept. 8, Capt. Anson Mills' command stumbled on a village at the Slim Buttes in northwestern South Dakota. Mills quickly captured the village and called for reinforcements. The infantry battalion came up the next day and Calhoun's company was posted on the camp's south side in case an attack came from that direction.

The famished soldiers lived off captured provisions until a train of supplies was sent from Deadwood. Slim Buttes was the army's first major victory of the Sioux War, and from the village numerous 7th Cavalry items were recaptured. Crook's column proceeded to the Black Hills, where it remained for several weeks to recuperate. While in the hills, photographer Stanley Morrow took his celebrated series of photos documenting the hard campaign. In a group photo of the infantry battalion officers, which Capt. Tobey aptly titled, "Starvation Crowd of 1876," Calhoun can be seen.[20]

On Oct. 24, the tired expedition marched into Camp Robinson and was thankfully disbanded. Calhoun and Company F were assigned to nearby Red Cloud Agency. Camp Robinson had originally been established to guard the agency and the various Oglala Sioux, Northern Cheyenne, and Arapaho bands who lived there. Calhoun was not overly impressed with his new station, once writing, "I cannot say that I particularly like this station, as it is one of the worst in the country."[21]

Lieutenant Frederic Calhoun, 1878
Courtesy Elizabeth Bonney

Shortly after Company F arrived, Tobey was detailed as the army Indian agent, and was frequently assisted by Calhoun. By early November, Calhoun completed a census of the people of Red Cloud, determining that its population totaled nearly 10,000 less than had been reported by the civilian agent the previous spring.[22] In December, he was granted a well-deserved two months leave.

After returning to his company in early February, Calhoun was sent in March to help Lt. William P. Clark enlist Indian scouts at Spotted Tail Agency. In May, he was detached to move supplies to troops operating in the Black Hills. At the same time, hundreds of northern warriors and their families began to return to the Nebraska agencies. Understandably, Calhoun remained bitter toward these surrendering Sioux and Cheyenne. In April, he wrote a friend on the apparent end of the Sioux War hostilities:

"This ends the Indian War for the present – Of Course it does not permanently end it for history will repeat itself and in a year or two they will all go out again. It is a very unsatisfactory ending for me, as I wanted to see them completely crushed, if not exterminated. Of course the massacre of the entire Indian race would not repay my loss of last summer, but the only way to insure lasting peace with them is to give them a good sound whipping, and these northern Indians have never yet been whipped."[23]

Constantly reminded of his loss, Calhoun also recovered his brother's watch from Indians near the agency, which he sent to Margaret.[24]

In the spring of 1877, Calhoun wanted to accompany the party sent to the battlefield to recover the remains of Custer, James Calhoun and the other officers. He wrote a friend, "I would like to be present when my brothers remains are removed and the Generals (Custer) widow and parents are anxious to have me attend to their people."[25] Regardless of his and the Custers' desires, Calhoun was not permitted to go with the recovery detail.

Later, Calhoun attended the reburial of James and the other officers at Fort Leavenworth National Cemetery. He left Camp Robinson on July 19, in charge of a detail escorting military prisoners to Omaha. He then joined Margaret, Mrs. Algernon Smith and Mrs. George Yates, as their husbands were buried with impressive ceremony on Aug. 3. Nearly 300 carriages were in the procession, including the band and two companies of the 23rd Infantry. Mrs. Gen. William T. Sherman was also in attendance. After the caskets were lowered into the ground "with loving hands, covered with a profusion of wreaths and boquets," the burial salute was fired. On Aug. 7, Calhoun returned to duty at Robinson.[26]

But the Indian Wars were not over for Calhoun, appointed post adjutant at Camp Robinson. In fact, a series of extraordinary events would take place that ironically would bring together famed war leader Crazy Horse and Fred Calhoun at the close of the Great Sioux War of 1876-77.

Crazy Horse and his band had surrendered May 6, 1877, and spent an uneasy summer in camp near the agency. As summer ended, rumors spread that he was planning to break away and renew the war with the whites. A decision was made to separate him from his people. On Sept. 4, Crazy Horse fled to Spotted Tail Agency, some 40 miles to the east. Here he was persuaded to return to Camp Robinson. Closely followed by army Indian scouts, an apprehensive Crazy Horse spent most of the next day on the return journey.

Just after 6 o'clock in the evening of Sept. 5, 1877, Crazy Horse was fatally wounded by a guard's bayonet in a scuffle in front of the post guardhouse. The guard was Private William Gentles, a member of Calhoun's Company F.[27] In a twist of history, the wounded Crazy Horse was taken into the adjutant's office – Calhoun's own office – in the building next door where he died about five hours later. One can only conclude that the Indian's death must have been sweet revenge for Calhoun for his brother's death at the Little Big Horn.

After more than a year at Camp Robinson, Calhoun's company returned to Camp Douglas, arriving Nov. 12, 1877. Calhoun spent some months on regular garrison duty, until the Bannack War broke out in the summer of 1878. In mid-June, three 14th Infantry companies, including Calhoun's, were sent north to Fort Hall Agency in Idaho, where they performed field service and guard duties. On one occasion, he was attached to a 5th Cavalry company scouting for roaming Bannacks still away from the reservation. Early in November, the infantrymen returned to Camp Douglas.

On Feb. 6, 1879, Calhoun began a three-month leave and returned to Monroe, where, on Feb. 20, he married Emma Reed, the daughter of David and Lydia Reed. She was the sister of Harry Armstrong Reed and the niece of George Custer. The Reed family hurriedly delivered the invitations on Saturday, the 15th, for the wedding ceremony to be held on Thursday, the 20th, in the Monroe Methodist Episcopalian Church. The bride looked "lovely" and the groom, "attired in full-dress uniform," made "quite a striking appearance." Calhoun officially entered the Custer family, though too late. After the wedding, the newlyweds took a train south to visit friends in Ohio and Indiana before returning to Fort Douglas.[28]

Emma, a favorite of the Custers, had also played a small part in the Little Big Horn saga. Elizabeth once wrote George that Emma was sweet, "but is quite fascinatingly different...gifted with the Custer trait-an inclination to coquette."[29] As with other friends and relatives, the Custers wanted her to visit Fort Lincoln, but that evidently did not come to pass until 1876 when Custer asked Emma to spend the summer with Elizabeth while the troops were in the field. Consequently, after he left Washington in early May, Custer stopped in Momoe to visit his parents and left the next day with Autie (Harry Armstrong) and Emma Reed.

When the 7th Cavalry left Fort Lincoln on May 17, Elizabeth and Margaret Calhoun, as historical accounts vividly document, accompanied the column for the first day's march. Emma Reed also may have gone along. Many years later, she wrote, "My aunts and I went one day's march with them in company of the paymaster; then, the men having been paid at a safe distance from temptation, we returned in the ambulance that had taken us on this little trip."[30]

Curiously, her presence on the trip was not noted by Mrs. Custer, and consequently it is a fact not generally known in the Custer literature. Her presence at the dramatic parting is only logical: a brother and uncle were on the expedition. Why should she not be afforded the privilege of a last visit? Weeks later, Emma was present at another poignant scene when army officers brought the dreadful news to the Custer house at Fort Lincoln. Afterward, she left with Elizabeth and Margaret for Monroe.

After returning to Fort Douglas, Fred Calhoun suffered an unfortunate mishap in the fall of 1879. That October, while riding through a narrow canyon, he was hit in the eye by a tree branch. The slow-healing injury seriously affected his vision. In April 1880, he traveled east on sick leave to seek treatment. By summer, under the care of a Cincinnati physician, he was slowly recovering the use of his eye. While on leave, the Calhouns spent much of their time in

Monroe. Although he finally returned to duty in September, the eye injury continued to bother him for years.[31]

Between 1881 and 1884, Calhoun's company was stationed at the Cantonment on the Uncompahgre in southwestern Colorado. This post was established in 1880 during the Ute War to serve as a supply base for troop operations in the region. In March and April 1882, he again went on leave to Monroe. In June 1884, the 14th Infantry was ordered to new posts in the Department of the Columbia; Company F was assigned to Vancouver Barracks, Wash. In late 1884, Calhoun again went on leave, returning to duty in January 1885.

At his new post Calhoun spent time on regular company duties and on special duty on the firing range for department competition. He occasionally took hunting leaves and temporarily commanded Company C in the absence of its regular officers. Also in the fall of 1885, Margaret Calhoun lived with Emma and him at Vancouver Barracks.[32]

The only field service for Calhoun at Vancouver came in late 1885 when violence between Chinese and white residents in Seattle threatened. A call was made for federal troops to aid civil authorities. In what was known as the "Puget Sound Expedition," the entire 14th regiment was dispatched from Vancouver Barracks, serving on this duty from Nov. 7 to 17. Answering a second call from Feb. 9 to 25, 1886, a battalion, including Calhoun's company, again was sent to

**Margaret Custer Calhoun
date unknown**
Courtesy Elizabeth Bonney

Seattle. Calhoun spent the balance of 1886 on regular garrison duty, except for short periods of general court martial duties at Forts Walla Walla and Coeur d'Alene and detached service at Fort Townsend to break the routine.

After a dozen years as a second lieutenant, Calhoun received his first lieutenancy in February 1887. With the promotion, he was transferred to Company C. That year he was ordered to the Infantry and Cavalry School of Application at Fort Leavenworth. But since the eye injury in 1879, Calhoun had experienced problems with his eyesight, occasionally suffering periods of double vision. So, he requested not to be sent to the school, fearing he would be unable to pass the course.33 Instead, he was sent on detached service to Fort Townsend on the Puget Sound, arriving Oct. 17, 1887.

An undated photograph of Frederic Calhoun, looking very worn
Courtesy Elizabeth Bonney

At Townsend, Calhoun was assigned temporary duty with Company A. While he was there, his physical condition deteriorated. For several years he had experienced difficulty with movement, and it became progressively more difficult for him to perform his military duties.

Since September 1888, he had been under almost constant observation by the post surgeon and was diagnosed with a disorder of the nervous system, which interfered with his leg movements. In August 1889, Calhoun returned to Vancouver Barracks and immediately went on leave. Finally on Dec. 6, he asked to go before a retirement board.[34] In January, he appeared before the board, presided over by Gen. John Gibbon.

The board's medical officers who were familiar with his problems concluded that Calhoun had been suffering from progressive spinal sclerosis since 1879. He had "a general in coordination of muscular movements with his eyes closed" and difficulty moving around at night. He also had attacks of neuralgia in the lower extremities, and could not sustain any physical or mental exertion. They felt his disability had developed slowly over the years and was by this time permanent. As a result, the board declared that Calhoun was "permanently disqualified for the performance of his military duties" because of disability originating in the line of duty.[35]

In February, Calhoun, who had been sick in quarters since November, was found incapacitated for active service and ordered home. He left Vancouver Barracks on March 7 on a leave of absence. On May 6, 1890, Calhoun was retired from the army "on account of disability incident to the Service." At that time he was reported as being in Springfield, Mo.[36] During the 1890s, the Calhouns apparently resided in Grand Rapids, Mich. Meanwhile, he became an occasional topic of conversation in the famous correspondence between Frederick Benteen and Theodore Goldin. Benteen reiterated the old complaints of the horse loan back on the Yellowstone Expedition and the Fort Lincoln petition. Referring to Calhoun's promotion several years before his retirement, he snidely remarked, "He got his first lieutenancy and soon retired (A something I look on as a kind of Swindling the Govnt) How does it strike you?"

In 1896, he pointed out that Lt. Henry Jackson had married a cousin (Elizabeth in 1870) of the Calhouns, creating yet another 7th Cavalry family line. Intermarriage between the Calhouns and the Custer family brought out the caustic comment, "they played things in general to suit themselves."[37]

By the turn of the century, the Calhouns had moved to Wellesley, Mass., where their only child, Emma May, entered Wellesley College. By this time Calhoun's mental state and physical health were failing, and he became more of a burden for his wife. In 1901, Elizabeth Custer, on behalf of Mrs. Calhoun, wrote to the surgeon general to see if he could be placed in a government hospital. Mrs. Custer related how Emma Calhoun "has suffered tortures in caring for him and borne it with true Custer courage," but now her husband was "becoming dangerous and it is necessary for her to place him in an institution." The surgeon general quickly replied that if proof of insanity could be provided, Calhoun could be admitted to the Government Hospital for the Insane in Washington, D. C. (present-day St. Elizabeth's Hospital).[38]

On March 20, 1904, Calhoun died at age 56 at Wellesley and was buried in Cincinnati. The cause of death was reported as "locomotor ataxia," also known as tabes dorsalis, "a syphilitic disorder of the nervous system marked by pain, lack of coordination of voluntary movements and reflexes, and disorders of sensation, nutrition and vision."[39]

Emma Reed, who survived her husband until her death in December 1949, was buried beside him in Cincinnati.[40]

Acknowledgements
The author wishes to thank Brian Pohanka, R. Eli Paul and Larry Vorderstrasse for assistance in preparing this article.

ENDNOTES

[1] Frost, Lawrence A., With Custer in '74 (Provo: Brigham Young University Press, 1979), xxii-xxiii; Monroe Record (Mich.), March 24, 1904.

[2] Exhibit B, "Statement of Mr. F. S. Calhoun," found with "Report of Examining Board," Calhoun's Appointment, Commissions, and Personal File (569 ACP 1875), RG 94, National Archives, Washington, D. c., hereafter cited as "Calhoun ACP File."

[3] Adjutant General's Office, Official Army Register of the Volunteer Force of the United States Army, Part V, Ohio, Michigan (Washington: Government Printing Office, 1865), p. 366.

[4] "Statement of Mr. F. S. Calhoun," Calhoun ACP File.

[5] Frost, Lawrence A., Custer's 7th Cavalry and the Campaign of 1873 (El Segundo: Upton & Sons 1985), p. 61, p. 178; John M. Carroll, ed., The Benteen-Goldin Letters on Custer and His Last Battle (Lincoln: University of Nebraska Press, 1991), p. 240.

[6] Custer, Elizabeth B., Boots and Saddles (New York: Harper & Bros., 1885), p. 128.

[7] Petition dated April 22, 1874, Calhoun ACP File.

[8] Carroll, Benteen-Goldin Letters, pp. 239-240.

[9] Carroll, John M., and Frost, Lawrence A, eds., Private Theodore Ewert's Diary of the Black Hills Expedition of 1874 (Poscataway, N. J.: CRI Books, 1976), p. 6.

[10] Telegram dated Dec. 19, 1874, Calhoun ACP File.

[11] Special Orders 283, Headquarters, Department of Dakota, Dec. 29, 1874, Calhoun ACP File.

[12] "Report of Examining Board in Case of Fred. S. Calhoun," Calhoun ACP File.

[13] Army and Navy Journal March 13 and 20, April 10 and July 17, 1875.

[14] Unless otherwise noted, all information on Calhoun's military assignments are compiled from the 14th Infantry Regimental Returns. Numerous articles in the Corinne Daily Mail report on the Bear River difficulties.

[15] Monaghan, Jay, Custer (Boston: Little, Brown & Co., 1959), p. 366.

[16] Meketa, Ray, ed., Marching with General Crook (Douglas, AK: Cheechako Press, 1983), Capron Diary, entries for July 8-9, 1876.

[17] See "Exact Copy of a Notebook Kept by Dr. V. T. McGillycuddy, M.D. While a Member of the Yellowstone and Big

Horn Expedition May 26-Dec. 13, 1876," Nebraska State Historical Society Archives, Lincoln, Neb., for the correct version. Incorrect versions published in Little Big Horn Associates Newsletter, May 1988 and November 1990 issues.

[18] The text of Moylan's letter is found in Frost, Lawrence A, General Custer's Libbie (Seattle: Superior Publishing Co., 1976), pp. 245-246.

[19] For a complete account of the Horsemeat March, see Greene, Jerome A, Slim Buttes. 1876, (Norman: University of Oklahoma Press, 1982).

[20] Hedren, Paul L., With Crook in the Black Hills (Boulder, Co.: Pruett Publishing Co., 1985), pp. 54-55.

[21] Calhoun letter dated April 27, 1877, Calhoun ACP File. For a complete history of Camp Robinson in the Sioux War, see Thomas R. Buecker," A History of Camp Robinson, Nebraska 1874-1878," M.A thesis, Chadron State College, Chadron, Neb., 1992.

[22] For an overall view of 1876-77 agency operations, see Thomas R. Buecker and R. Eli Paul, eds., The Crazy Horse Surrender Ledger (Lincoln: Nebraska State Historical Society, 1994).

[23] Calhoun letter dated April 27, 1877, Calhoun ACP File.

[24] Carroll, John M., Cyclorama of Gen. Custer's Last Fight (El Segundo: Upton and Sons, 1988), p. 54.

[25] Calhoun letter dated April 27, 1877, Calhoun ACPFile.

[26] Army and Navy Journal, Aug. 11, 1877.

[27] William Gentles died at Fort Douglas on May 20, 1878, and is buried in the national cemetery there.

[28] Monroe Commercial, Feb. 21, 1879.

[29] Merrington, Marguerite, ed., The Custer Story (New York: The Devin-Adair Co., 1950), p. 251.

[30] "The Observer" column, Monroe Commercial, Jan. 7, 1950.

[31] Calhoun letter from Madison, Ind., July 5, 1880, Calhoun ACP File.

[32] Frost, in General Custer's Libbie, makes the mistake of placing Vancouver Barracks in Wyoming (p. 259). Shirley Leckie in her excellent biography, Elizabeth Bacon Custer and the Making of a Myth (Norman: University of Oklahoma Press, 1993), p. 241, unfortunately repeats the error.

[33] Calhoun letter, dated May 21, 1887, Calhoun ACPFile.

[34] "Report of the Medical Officers of the Retiring Board," Jan. 16, 1890, Calhoun ACP File; Letter, Calhoun to the Adjutant General, Dec. 6, 1889, Calhoun ACP File.

[35] Army and Navy Journal, Jan. 11, 1890; "Report of Medical Officers," Calhoun ACP file.

[36] Army and Navy Journal, Feb. 15, May 10, 1890.

[37] Carroll, The Benteen-Goldin Letters, pp. 236, 243, 240.

[38] Elizabeth Custer to Surgeon General George M. Sternberg, Oct. 8, 1901, Calhoun ACP File. Neither his ACP File nor pension records indicate that he was ever admitted.

[39] Monroe Democrat March 25, 1904: Army and Navy Journal, March 26, 1901; Webster's Ninth Collegiate Dictionary (Springfield, Mass.: Merriam-Webster Inc., 1985), pp. 701, 1199.

[40] "The Observer" column, Monroe Commercial, Jan. 7, Aug. 12, 1950.

History of the 14th Infantry Regiment 1865 to ca. 1896

By Colonel Thomas Anderson[1]

FRONTIER SERVICE.

In some way it became known before the order was issued that the 14th Infantry would be designated for a tour of duty on the Pacific Coast.

After the disbanding of the volunteer forces many wild characters found their way into the ranks of all the Regular regiments. Some of these men had done good service in the field, but they adopted a theory that as the War was over, discipline would be relaxed and that they should be permitted to have what they were pleased to call "a high old time." Nor was this pleasing theory confined to the ranks; a number of officers came to grief from practices under an epicurean philosophy which the War Department deemed "more honored in the breach than in observance." Thus it happened that the 14th got more than its share of Bacchanalian warriors.

In the last week of July the 2d Battalion left Richmond for New York City, followed in a few days by the 1st. Both assembled at Hart's Island, where they made their preparations for a trip to California via Panama. From the 2d Battalion alone, 221 men deserted in two weeks. They were all reported as bounty jumpers, assigned just before the close of the War.

It sailed from New York City on August 15, 1865, under Major Louis H. Marshall. This officer only reported for duty a few days before, having been on staff duty as colonel, A. D. C., up to the 28th of the preceding July. In passing over the Isthmus, the new men gave proof of their quality, for they proposed to take Aspinwall and Panama, and it was only by the courageous and forcible efforts of the officers, non-com. officers and old soldiers that the unruly

---

[1] Excerpt from regimental history written by Colonel Thomas Anderson, see Rodenbough and Haskin, 1896, in the bibliography.

element was subdued and the battalion safely embarked on the Pacific side.

Col. Chas. S. Lovell, who had been promoted to the colonelcy of the regiment upon the retirement of General Paul on February 16, 1865, reported for duty at Hart's Island, N. Y. H., August 28, 1865. He was the first full colonel to assume command of the regiment since its reorganization. The organization of the Third Battalion was begun and vigorously pressed. At the same time the First Battalion was filled up, and on October 16th the field, staff and band of the regiment and four companies of the 1st Battalion, E, F, G and H, under Colonel Lovell, left New York and landed in San Francisco, November 12th, taking station temporarily at the Presidio. Cos. A, B, C and D followed two weeks later.

The Third Battalion, under Major Chapin, followed in November, arriving at San Francisco early in December. Here there was an outburst of turbulent hilarity which manifested itself chiefly in cutting off the pigtails of the Pagans. The battalion was hurried away to Arizona, where the exuberance of the young warriors could find less objectionable play in cutting off the scalp-locks of Apaches. The headquarters of the battalion under Major Chapin was fixed at Goodwin, with companies detached to Crittenden, Lowell, Grant and Bowie.

In October of 1865, the Second Battalion, under Major Marshall, had been sent to the Department of the Columbia, the officers for duty being Captains Ross, Coppinger, O'Beirne and Walker, and Lieutenants Henton, McKibbin, Wharton, Porter, Perry, Collins, Tobey and Kistler. Colonel Lovell soon followed with his regimental staff, Downey and Bainbridge, establishing headquarters at Fort Vancouver, December 8th.

In January of 1866, the 1st Battalion, under Major Hudson, was ordered to Drum Barracks and from thence to Fort Yuma, California, at which post the headquarters of the battalion was established February 6th, Co.'s A, B, C, G and H constituting the infantry garrison, Co.'s E and F having been left at Drum Barracks, and Co. D sent to Date Creek. On the 17th Captain O'Connell succeeded to the command. Subsequently Co. H was sent to Date

Creek, and B and D to McDowell. In October the headquarters of the battalion were at Fort Whipple with Captain Krause in command.

The headquarters of the regiment remained at Vancouver Barracks until June, 1866, when it was ordered to San Francisco and thence to Arizona, where it was established September 6, 1866. The band was left at Fort Yuma.

In January, 1867, the headquarters of the regiment was transferred to Camp Lowell, Tucson, Arizona, where January 23, 1867, the provision of the act of Congress of July 28, 1866, altering the battalion organization into a regimental one was carried out and the 1st Battalion of the regiment with two companies subsequently added, became the 14th Infantry.

The 2d Battalion, which had remained in Oregon and Washington, became the 23d Infantry, and the 3d Battalion, which was serving in Arizona, became the 32d Infantry. On the 16th of April the headquarters of the regiment were established at Fort Yuma, in which military Tophet it remained until May, 1869.

Under the reorganization of 1866, the captains were distributed as follows: To the 14th Infantry, Captains Ilges, Smedberg, Krause, Wharton, Weir, Van Derslice, Bainbridge and Vernou. To these were added Captains Hamilton and Davis for the two additional companies.

Captains D. B. McKibbin, Brown, O'Beirne, Downey, Miller, Perry and Fergus, were assigned to the 32d, and Captains Ross, Clay, Coppinger, Brady, Walker, Sinclair, Henton and Browning were assigned to the 23d Infantry.

Of the field officers the 14th retained Colonel Lovell and Lieutenant-Colonel Wallen; Maj. L. H. Marshall went to the 23d and Major Chapin to the 32d. In January of 1867, the 14th Infantry was distributed at the following stations: Yuma, McDowell, Mojave, Lincoln and Camp in Skull Valley, without question the worst in the country. During this tour of duty nearly every monthly return contains a record of Indian scouts; some months nearly every company would be out. In September, 1868, the distance marched by these scouting parties aggregated 1000 miles, equivalent to double the distance elsewhere. Two companies marched 350 miles in August. The skirmishes rarely rose to the dignity of a battle, but they taxed the courage and skill of the participants to the utmost. One of the commonest entries is that of "mail carriers killed by Indians." Several hundreds of miles of wagon road were made by the regiment, and when the men were in camp they were almost constantly engaged in building barracks and quarters.

In the reorganization of the Army in 1869, the 45th Infantry, one of the Veteran Reserve regiments, was consolidated with the 14th Infantry. In compliance with S. O. No. 17, A. G. O. 1869, the 14th Infantry was transferred to Nashville, Tenn., the headquarters of the 45th Infantry, taking with them the officers, non-commissioned officers and ten men of each company. The other enlisted men were discharged or transferred to other regiments remaining in the Department of Arizona. The consolidation was carried out, the result appearing in the monthly return for July. The field officers assigned to it were Col. C. S. Lovell, Lieut.-Col. Geo. A. Woodward and Maj. M. M. Blunt, Lieutenant McCammon was made adjutant and Lieutenant Steele was retained as quartermaster.

The captains of the reorganized regiments were: Ilges, Krause, Van Derslice, Freudenberg, Trotter, Hamilton, Bainbridge, Carpenter, Burke and Davis. Their stations were Nashville, Humboldt, Chattanooga, Louisville, Jeffersonville, Lebanon and Union, W. Va.

In April, 1870, the regiment was transferred to Fort Randall, Dakota, on account of a threatened Indian war. In August it was transferred to the Department of the Platte, with headquarters at Fort Sedgwick, the regiment and post being under Lieut.-Col. G. A. Woodward. In the following March (1871) the headquarters was transferred to Fort Laramie, Wyo., where General John E. Smith reported and assumed command. Colonel Lovell had been retired December 15, 1870. General Gordon Granger, a colonel unassigned, was assigned to the regiment, vice Lovell, but on the 20th of December General Smith, who had been assigned to the 15th Infantry, was transferred to the 14th, General Granger at the same time being assigned to the 15th Infantry. Colonel Lovell died very soon after his retirement. He was loved and respected by the regiment. He was sincere, courteous and just, a good soldier and a good friend. The new colonel was a very different man. From all accounts of him he knew little and cared less for the traditions of the Service. He was a rough and ready fighter, who had done good service as a volunteer general. He would have led his regiment into a fight as gaily as into a frolic, but opportunity was never given him.

In February, 1874, Lieutenant L. H. Robinson was killed in an Indian fight near Laramie Peak, while guarding a supply train. In the following August the regiment went to Utah, with headquarters at Fort Douglas. Four companies went on to Fort Cameron under Lieutenant-Colonel Woodward.

While this battalion was at Cameron, the Mormon Bishop John D. Lee was arrested and held there as a prisoner, pending his trial as the leader of the band of Danites (or destroying angels) who perpetrated the Mountain Meadow massacre. After his conviction he had his choice under the laws of Utah, as to whether he should be hung, beheaded, or shot. He chose the latter method of execution. To carry out the rules of poetic as well as moral justice he was taken to the scene of the massacre and shot to death by musketry in March, 1879. A detachment under Lieutenant Patterson was sent down to preserve order. An attempt was made to convert Lee from the error of his ways, while he was confined at Cameron, but he maintained the scriptural doctrine to the last, "that the enemies of God should be exterminated root and branch," and finally met his fate with the equanimity of a martyr.

In 1876 the Sioux War broke out which opened up with the Custer massacre and the repulse of General Crook at the Rose Bud. In June, companies C, B, F and I (Burke, Kennington, Tobey, Murphy, Taylor, Yeatman, Calhoun and Lloyd), were sent to join Crook's column.

At Fetterman they met detachments from the 4th and 9th Infantry. The infantry column was placed under the command of Major Alexander Chambers, 4th Infantry, and hastening to join General Crook on the Little Goose Creek, enabled him to assume the offensive. Their only battle was at Slim Buttes, September 9th, where twenty-seven Indians were killed.

This column marched in three months 1139 miles. It was on the march from the Little Missouri to the Black Hills that the whole column was nearly reduced to starvation. Another company on escort duty marched 377 miles in one month. In November Companies D and G, under Captain Krause, were in (Crook's) the Powder River campaign, and were with McKenzie at the battle of Crazy Woman's Fork, November 26th, coming up with the infantry under General Crook. This column marched 735 miles. The officers present were Krause, Van Derslice, Hasson, Austin and Kimball. In 1877 one company was in the Nez Percé campaign and five under Major Bryant in the Bannock War, but they did not have a battle. Three companies, Trotter's, Krause's and Van Derslice's, were out the next year after the Bannocks.

In 1879 four companies, E, 1, H and K, under Trotter, Carpenter, McConihe and Taylor, and Major Bryant commanding, were hurried down to the scene of the Thornburgh massacre, but arrived too late to get into the battle. But they did have all the hardships and privations of a hard Indian campaign.

In all the Indian campaigns of the regiment, their endurance, patience, vigilance and bravery were tested to the utmost. They suffered from the most suffocating heat in Arizona and the most intense cold in Wyoming.

The Apaches and the Sioux were formidable enemies, but they dreaded them less than sand storms and snow storms, scarcity of food and bad water. Many men broke down under these trials, who easily endured all the hardships of the Rebellion.

Besides the battles mentioned in the narrative, detachments of the regiment were engaged in the following skirmishes:

February 23, 1866, Captain Walker and Lieut. T. F. Tobey with a detachment of fifteen soldiers of the 14th Infantry and twelve Oregon Volunteers, attacked and defeated a band of Snake Indians on Jordan Creek, Oregon, killing 18 and wounding 2 Indians. One man of the 14th was killed and 1 wounded.

On October 10, 1867, Captain Krause with a detachment of twenty-five men of the regiment attacked a Rancherio, twenty-five miles from Camp Lincoln, defeating the Indians, killing and wounding a number and capturing a lot of arms.

In a fight near Aqua Frio Springs, Arizona, November 13, 1867, Lieut. A. J. Converse and two men of Company C were wounded. Indians repulsed.

April 27, 1867, Lieutenant Western, with a detachment of ten men from Camp Logan, attacked a band of forty-five hostile Indians on Silvies River, fording the river neck deep. The Indians were defeated, 6 killed and a number drowned in trying to escape. Thirty-two horses and large amounts of supplies were taken. Complimented in orders (G. O. No. 32 Department Col. 1867).

Lieutenant Hasson, in the months of September, October, November and December, 1867, in command of detachments from his post, had engagements with the Apaches at Three Buttes, Hualopais Valley, Hitchie Springs and the Willows.

March 25, 1868, Captain Ilges and eight men attacked fifty Indians with stolen cattle at Cottonwood Springs, Arizona. The engagement lasted twenty minutes. Private Logan, Company B, was wounded. One Indian was killed and two wounded.

February 27, 1869, in an attack made by Apaches on a train near Camp Grant, Arizona, two men were severely wounded, but the attack was repulsed.

May 6, 1869, in an attack on a train near Grief Hill, one private of the regiment was killed, but the Indians were so impressed by the operations of breech-loaders, then used on them for the first time, that they regularly stampeded.

In May, 1881, Colonel Smith was retired and was succeeded by Lewis Cass Hunt, who was colonel of the regiment until his death, September 6, 1886.

In August, 1881, the headquarters of the regiment was transferred from Camp Douglass, Utah, to White River, Col., and in May 1883, they were removed to Fort Sidney, Neb., and in July 1884, to Vancouver Barracks, W. T.

In this department the regiment has had only the ordinary routine duty to perform, except the suppression of the anti-Chinese riots in Seattle in November 1885 and February 1886.

In September of this year Colonel Anderson was promoted to the Colonelcy of the regiment *vice* General Hunt. Lieutenant-Colonel Woodward was promoted to the colonelcy of the 15th Infantry on January 10, 1876. Lieut.-Col. Henry Douglas was promoted in his place on that date; he was promoted colonel of the 10th Infantry, July 1, 1888, and was succeeded by Lieut.-Col. I. D. DeRussy. Major M. M. Blunt was promoted October 4, 1874, lieutenant-colonel of 25th Infantry and was succeeded as major by Major Montgomery Bryant, who held the position until June 1882, When he was succeeded by Major W. F. Drum, who in his turn was promoted December 8, 1886, and was succeeded by Major Charles A. Wikoff, the present major of the regiment.

The regiment has as it stands to-day, twenty officers with war records, not counting those who have since served in Indian wars, nearly all of whom have been wounded in battle. Many of our "comrades and companions" have returned to civil life and are working honorably and successfully in civil pursuits. But the grave has closed over most of our men of '61.

"The brightest have gone before us
The dullest remain behind."

Nevertheless, those who remain, cherish the hope that those who succeed us may be encouraged by this history to do what the men of the 14th Infantry have always tried to do THEIR DUTY.

Elements of the 14th Infantry Regiment on parade at the Vancouver Barracks circa 1903
Courtesy Vancouver Barracks Military Association

ROSTER OF COMMISSIONED OFFICERS, 14TH INFANTRY[2]

Colonel, THOMAS M. ANDERSON.

Lieutenant-Colonel, I. D. DE Russy.

Major, CHARLES A. WIKOFF.

Adjutant, 1st Lieut. R. T. YEATMAN.

Quartermaster, 1st Lieut. J. H. GUSTIN

Co A   Captain A. H. BAINBRIDGE, 1st Lieut. G. T. T. PATTERSON, 2d Lieut. W. B. REYNOLDS.

Co B   Captain P. HASSON, 1st Lieut. J. MURPHY, 2d Lieut. J. P. O'NEIL.

Co C   Captain D. W. BURKE, 1st Lieut. WM. W. McCAMMON, 2d Lieut. E. T. WINSTON.

Co D   Captain C. B. WESTERN, 1st Lieut. F. S. CALHOUN[3], 2d Lieut. H. C. CABELL, JR.

Co E   Captain F. E. TROTTER, 1st Lieut. J. A. BUCHANAN, 2d Lieut. F. F. EASTMAN.

Co F   Captain T. F. TOBEY, 1st Lieut. C. A. JOHNSON, 2d Lieut. C. H. MARTIN.

Co G   Captain C. H. WARRENS, 1st Lieut. W. P. GOODWIN, 2d Lieut. W. A. KIMBALL.

Co H   Captain S. McCONIHE, 1st Lieut. S. J. MULHALL, 2d Lieut. W. R. SAMPLE.

Co I   Captain G. W. DAVIS, 1st Lieut. F. TAYLOR, 2d Lieut. A. HASBROUCK, JR.

Co K   Captain G. S. CARPENTER, 1st Lieut. R. A. LOVELL, 2d Lieut. W. K. JONES.

---

[2] Information reproduced from table in the original book.

[3] Fred Calhoun is still listed as a serving officer, so this narrative must have been written before 1890.

Custer's Other Brother-in-Law

Timeline of Lieutenant Frederic Calhoun's Career[4]

Apr. 9, 1847	Frederic S. Calhoun born.
May 12 – Aug. 19, 1864	Fred Calhoun enlists as a private with Company B, 137th Ohio Infantry.
1867 – 1869	Sometime in this period, Fred Calhoun follows his brother James Calhoun to the Pacific Northwest, as a civilian clerk.
1873	Fred Calhoun joins Lieut. Col. George Custer's Yellowstone Expedition as a civilian employee.
1873 – 1874	Fred Calhoun winters over at Fort Abraham Lincoln with his brother James and Margaret Custer-Calhoun, his wife, becoming one of the Custer Clique.
1874 – 1875	Fred Calhoun applies for a direct commission in the 7th Cavalry. He remains at or near Fort Abraham Lincoln through early 1875.
Mar. 10, 1875	Fred Calhoun receives commission as second lieutenant in the US Army and is assigned to the 14th Infantry Regiment.
Apr. 26, 1875	Fred Calhoun leaves Fort Abraham Lincoln.
May 19 – Jun. 12, 1875	Temporary Duty at Newport Barracks, Ky.
Jul. 9, 1875	Reported for duty with 14th Infantry Regiment at Fort Douglas, Utah. Later assigned to Company F, 14th Infantry Regiment.
1876 to 1877	**The Great Sioux War, or the Big Horn & Yellowstone Expedition**
Jun. 24, 1876	Fred Calhoun, along with Companies C, B, F, and I Camp Douglas, Utah and join expedition in the field.
Oct. 24, 1876	Calhoun's unit with the Big Horn & Yellowstone Expeditionary Force disbanded.

---

[4] From Buecker, 1994 and Keever, 1890.

Custer's Other Brother-in-Law

Oct. 30 - Dec. 5, 1876	Fred Calhoun and Company F are assigned to Camp Robinson, NE, and the Red Cloud Indian Agency.
Aug. 3, 1876	Fred Calhoun attends formal funeral ceremony of 7th Cavalry dead.
Aug. 12, 1876	Fred Calhoun appointed Post Adjutant of Camp Robinson, NE.
Sep. 5, 1876	Crazy Horse dies in Adjutant's Office at Camp Robinson, NE.
Nov. 4, 1877 – Jun. 7, 1878	Fred Calhoun, and Company F, reassigned to Camp Douglas, UT.
Jun. to Aug. 1878	**The Bannack War**
Jul. 5, 1878	Companies F, G & I, 14th Infantry Regiment ordered to Fort Hall, IT, supporting the Bannack War. Fred Calhoun later detached to command Company H, 5th Cavalry.
Nov. 4, 1878 - Apr. 10, 1880	Fred Calhoun and Company F, 14th Infantry Regiment at Fort Douglas, UT.
1879 to 1880	**The Ute War, or the White River War**
Feb. 5, – Apr. 23, 1879	Fred Calhoun on leave, marries Emma Reed, niece of George Armstrong Custer on Apr 20, 1879.
Oct. 1879	Fred Calhoun suffers and eye injury that will affect him for many years.
Apr .11, – Sep. 6, 1880	Fred Calhoun on sick leave for eye injury.
Sep. 7, 1880 – Aug. 8, 1881	Fred Calhoun returned with Company F, to Fort Douglas, UT.
Aug. 9 – Oct. 25, 1881	Fred Calhoun and unit assigned to Fort, Lyon, CO.
Oct. 30, 1881 – Jul. 21, 1884	Fred Calhoun and Company F arrive at Cantonment on the Uncompahgre, CO.

Custer's Other Brother-in-Law

Mar. 1 – May 31, 1882	On leave.
Jun. 1884	The 14th Infantry Regiment ordered to new posts in the Department of the Columbia.
Jul. 29, 1884 – Oct. 18, 1887	Fred Calhoun and his unit are stationed at the Vancouver Barracks, WT.
Fall 1885	Margaret Custer-Calhoun stays with the Calhouns.
Nov. 7 – 17, 1885	The 14th Infantry Regiment sent to Seattle, WT, to quell anti-Chinese riots.
Feb. 9 – 25, 1886	One battalion (including Fred Calhoun), return to Seattle, WT, to quell anti-Chinese riots.
Dec. 3, 1886	Fred Calhoun promoted to first lieutenant and transferred to Company C, 14th Infantry Regiment.
Oct. 19, 1887 – Aug. 27, 1889	Fred Calhoun transferred to Fort Townsend, WT for duty with Company A, 14th Infantry Regiment. These are the last set of orders posted into Fred Calhoun's book. There are other entries in later pages, but this is the latest dated document.
Aug. 28 – Nov. 27, 1889	Fred Calhoun on leave
Jan. 1890	Fred Calhoun appears at medical retirement board.
May 6, 1890	Fred Calhoun granted medical retirement from the army, at age 43.
Mar. 20, 1904	Fred Calhoun dies at age 56.

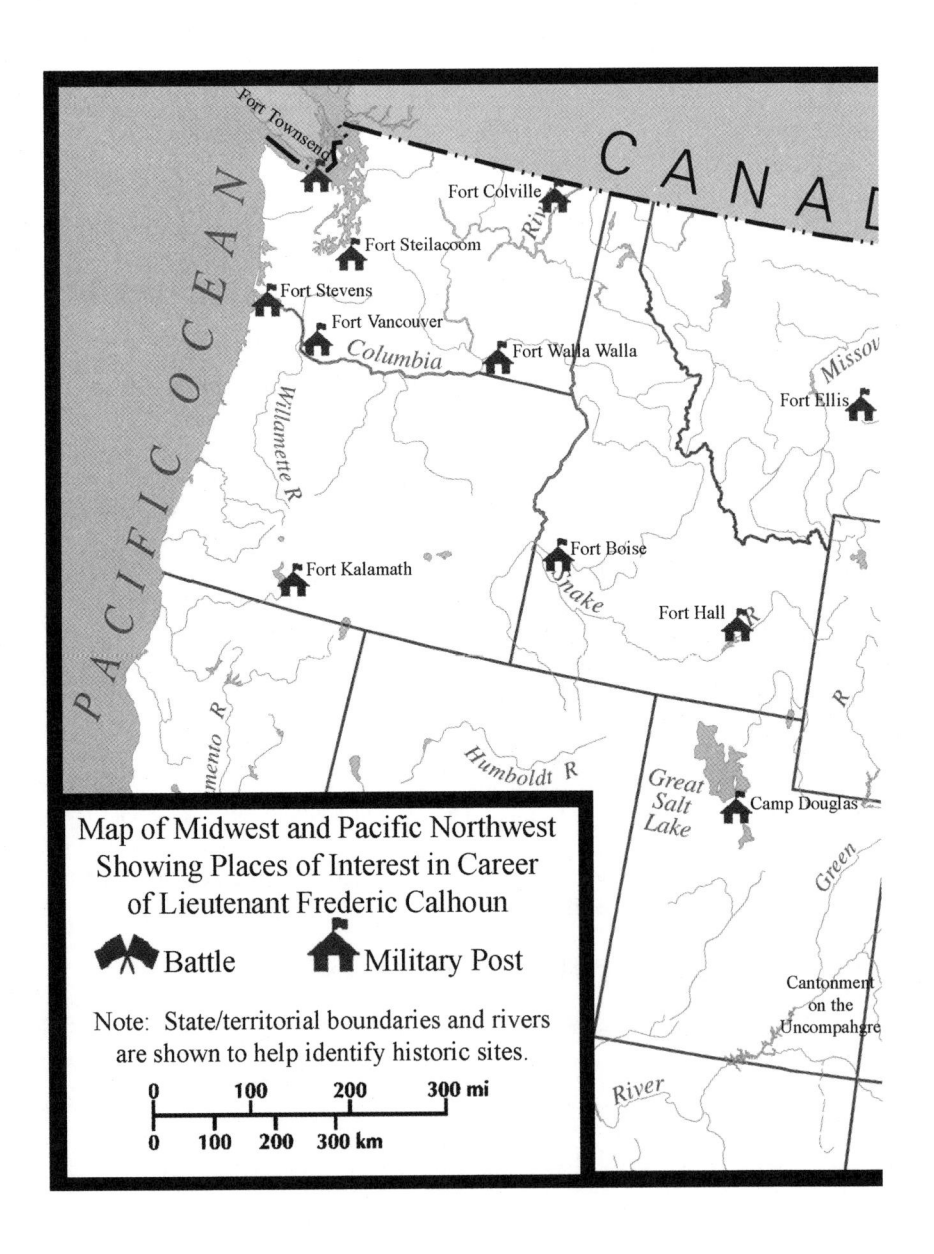

Map of Midwest and Pacific Northwest Showing Places of Interest in Career of Lieutenant Frederic Calhoun

⚑ Battle   🏠 Military Post

Note: State/territorial boundaries and rivers are shown to help identify historic sites.

Custer's Other Brother-in-Law

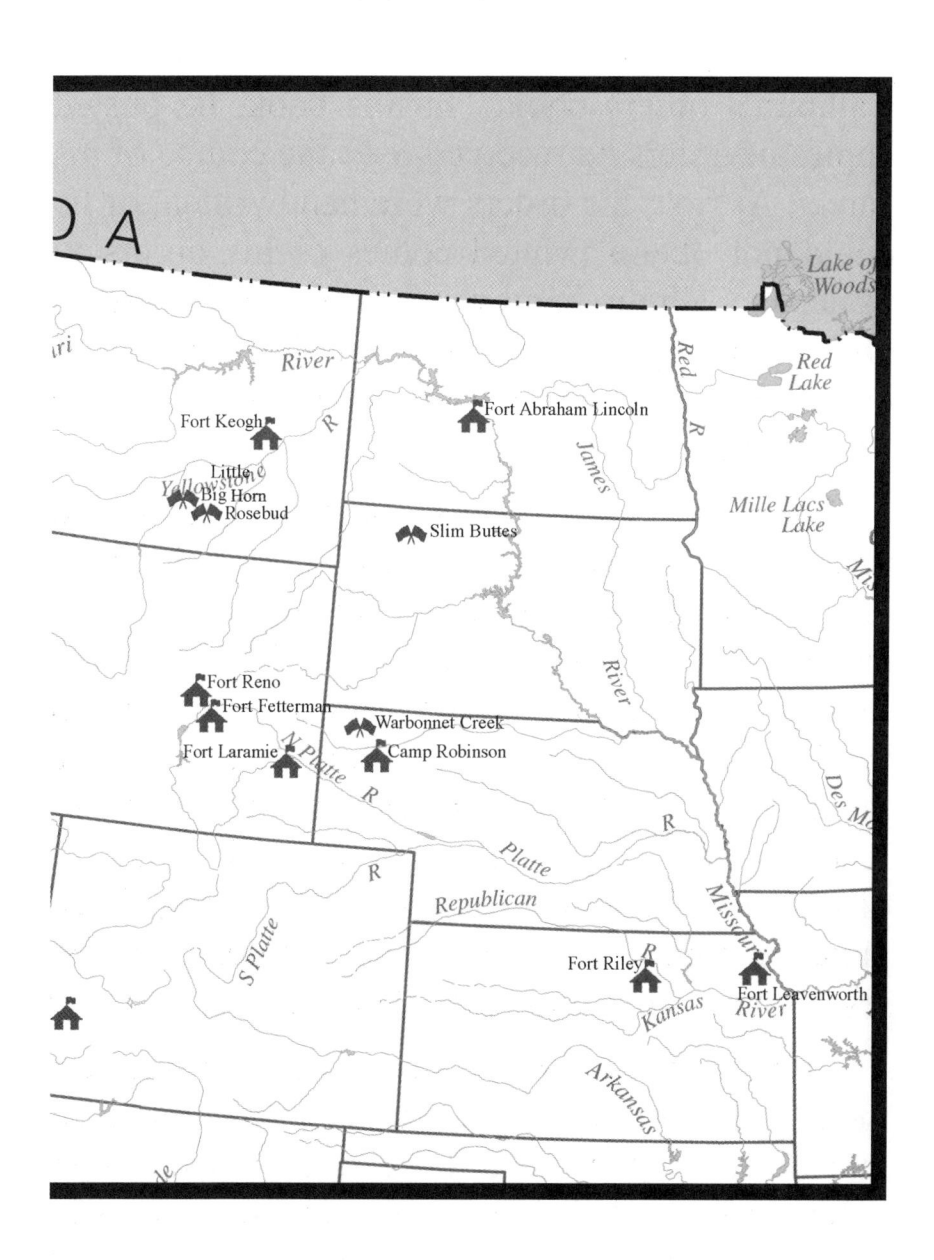

Using this Appendix

This appendix contains the contents of Fred Calhoun's orders book. In this book, he posted copies of orders he received over the course of his career. At first, his orders were handwritten, or he could not obtain printed copies of his orders to paste, so Calhoun hand wrote copies of his orders. Over time, typewriters and multiple copies of orders became available, and he posted these printed orders.

To make this book understandable to researchers, where the handwriting is hard to decipher, or the printed orders are too small to read easily, we have provided translations. When looking at this appendix, the original orders are on the left page, while a typed translation is on the facing page. In the case of easy to understand or read orders, only the original book page is displayed. His ledger book came with preprinted page numbers in the upper corners, which researchers can use to cross check dates.

Unlike a diary, Fred Calhoun did not interject many of his personal observations or comments in this book. When he did, he used red or other colored pencil. This is pointed out by the editors in their comments, which are contained in [brackets.]

Appendix A

Orders Book

of

Lieutenant Frederic S. Calhoun

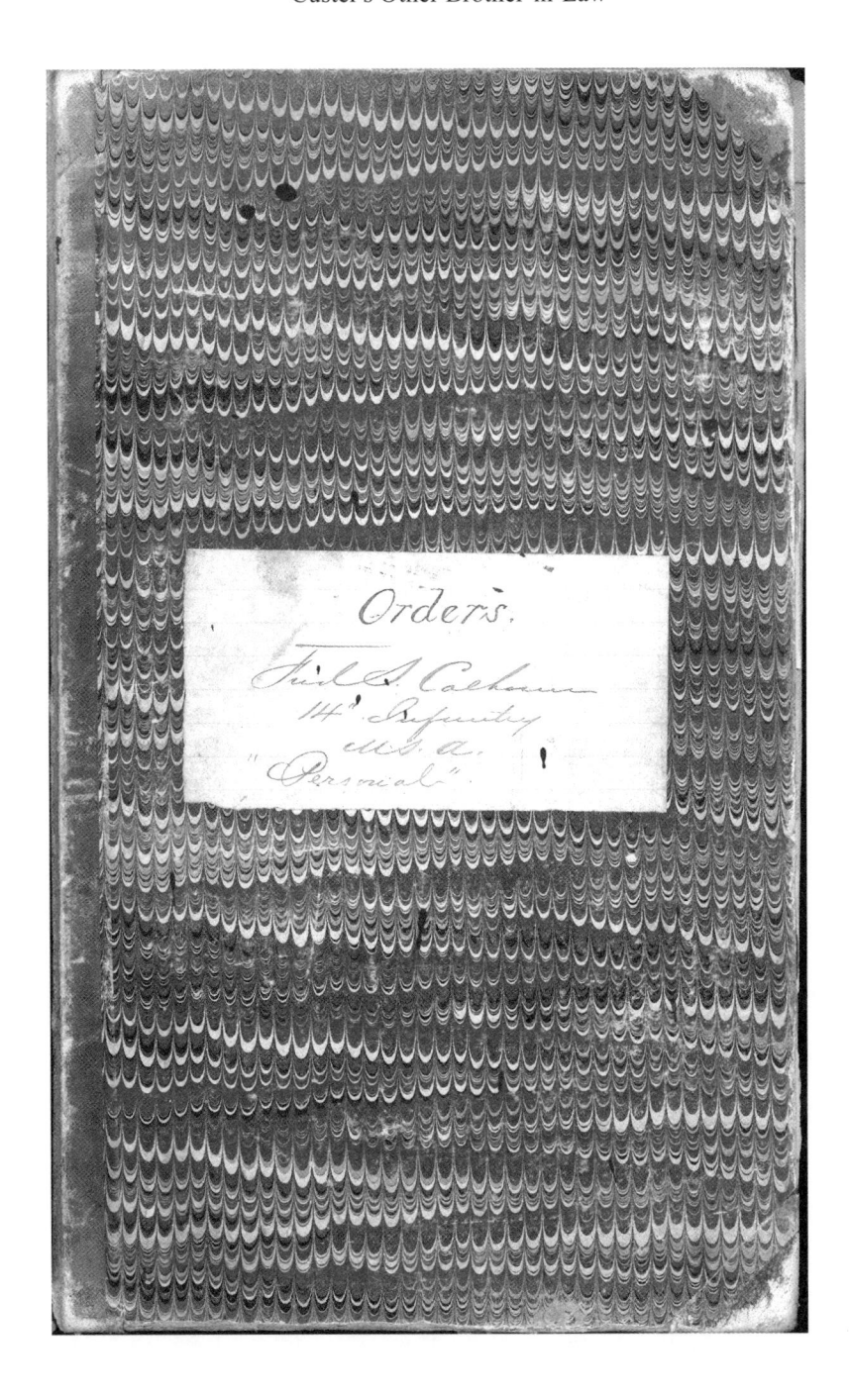

Orders.

Fred S. Calhoun
14th Infantry
U.S.A.
"Personal"

Orders.

Fred S. Calhoun

14th Infantry

U.S.A.

"Personal"

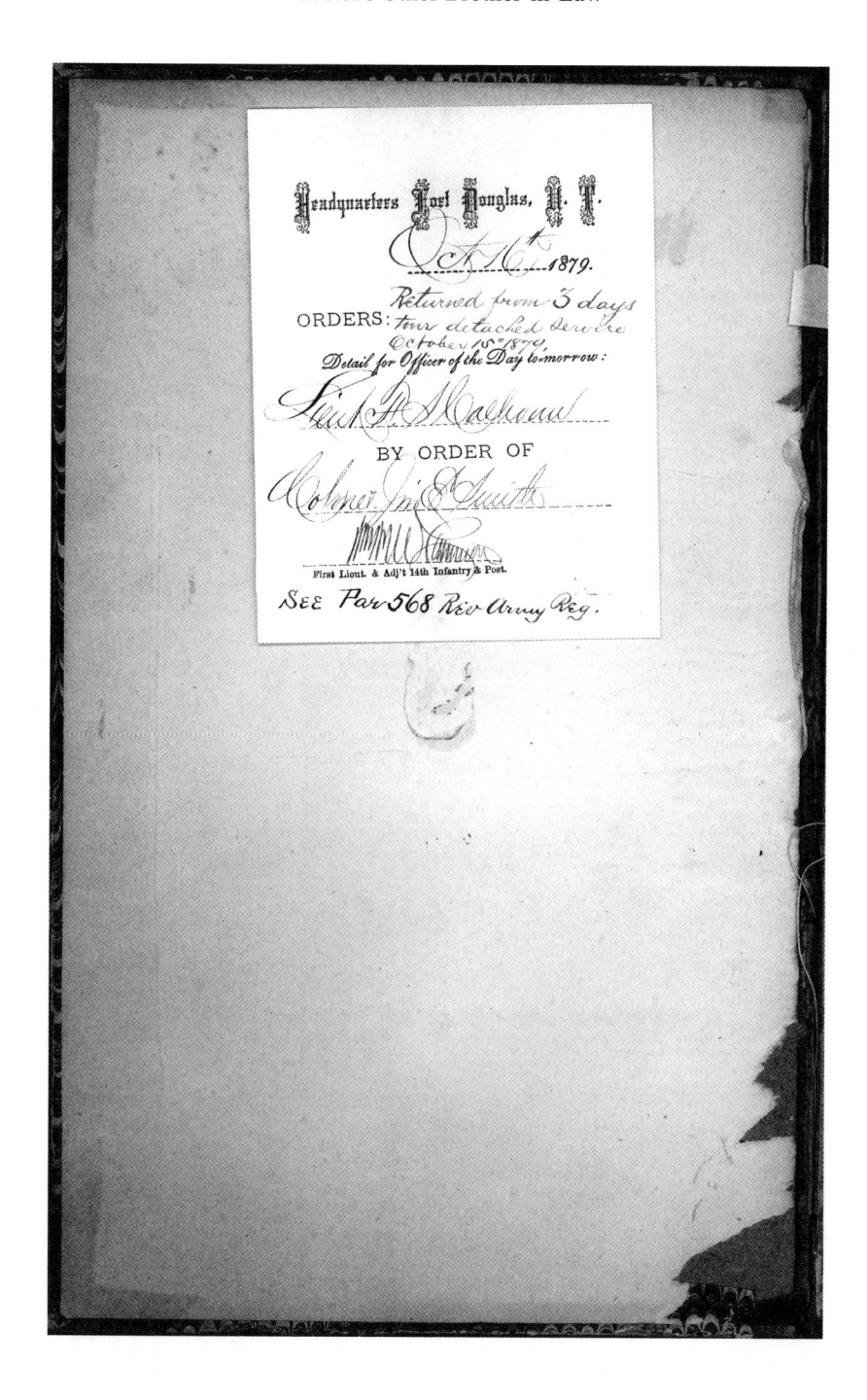

Headquarters Fort Douglas, U.T.
Oct 16th 1879.

ORDERS: Returned from 3 days tour detached service October 15th 1879.
Detail for Officer of the Day to-morrow:

Lieut F. A. Calhoun

BY ORDER OF

Colonel Jno. E. Smith

[signature]
First Lieut. & Adj't 14th Infantry & Post.

SEE Par 568 Rev Army Reg.

[Affixed label:]
Headquarters Fort Douglas, U.T.
October 16th 1879
ORDERS:
Returned from 3 days
tour detached service
October 15th 1879.
Detail for Officer of the Day tomorrow:
Lieut. F. S. Calhoun
BY ORDER OF
Colonel Jm E. Smith
William McCammon
[With McCammon's signature]
First Lieut. & Adj't 14th Infantry & Post
See Par 568 Rev Army Reg.

Custer's Other Brother-in-Law

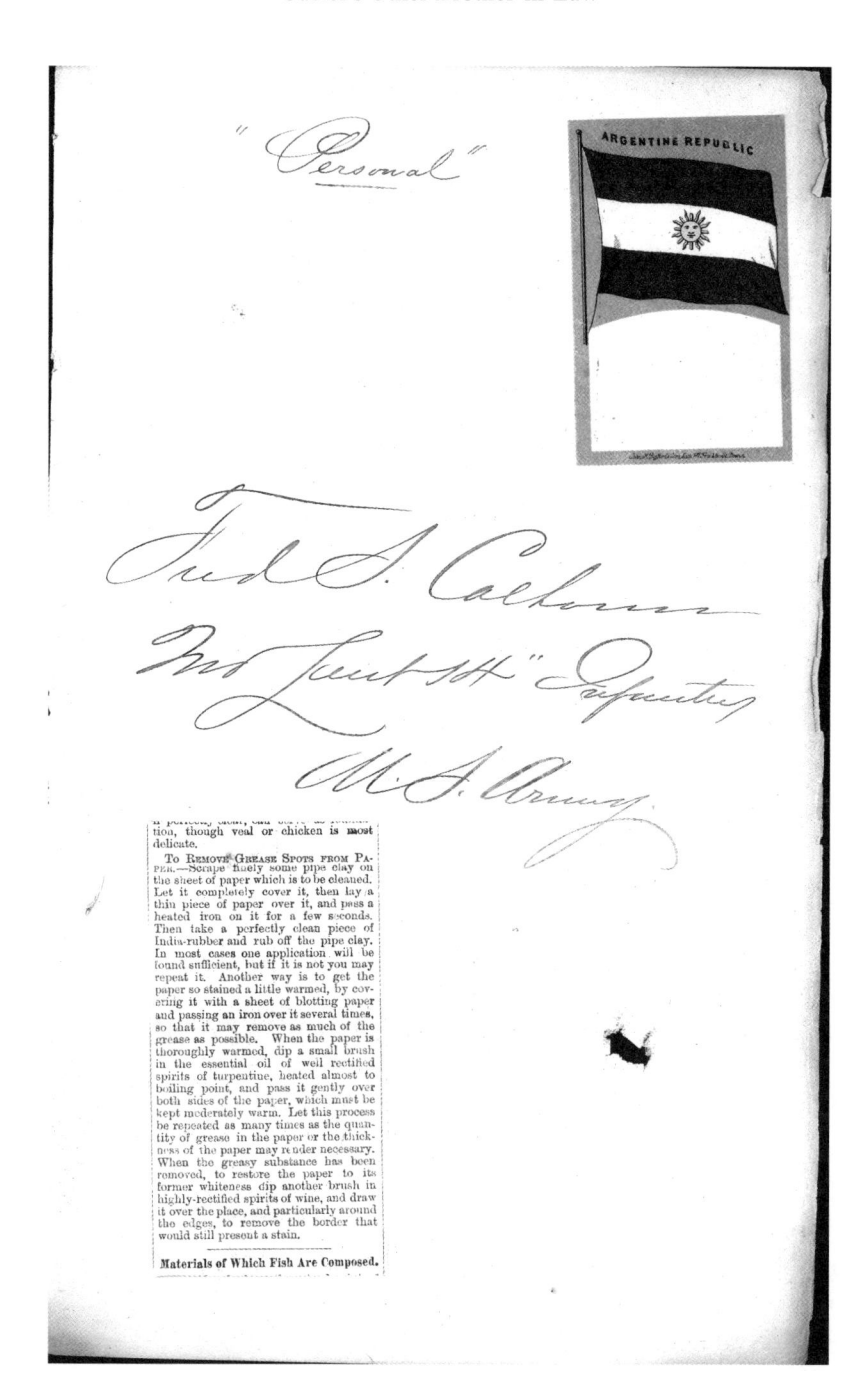

"Personal"

ARGENTINE REPUBLIC

Fred S. Calhoun
2nd Lieut 14th Infantry
U.S. Army

tion, though veal or chicken is most delicate.

TO REMOVE GREASE SPOTS FROM PAPER.—Scrape finely some pipe clay on the sheet of paper which is to be cleaned. Let it completely cover it, then lay a thin piece of paper over it, and pass a heated iron on it for a few seconds. Then take a perfectly clean piece of India-rubber and rub off the pipe clay. In most cases one application will be found sufficient, but if it is not you may repeat it. Another way is to get the paper so stained a little warmed, by covering it with a sheet of blotting paper and passing an iron over it several times, so that it may remove as much of the grease as possible. When the paper is thoroughly warmed, dip a small brush in the essential oil of well rectified spirits of turpentine, heated almost to boiling point, and pass it gently over both sides of the paper, which must be kept moderately warm. Let this process be repeated as many times as the quantity of grease in the paper or the thickness of the paper may render necessary. When the greasy substance has been removed, to restore the paper to its former whiteness dip another brush in highly-rectified spirits of wine, and draw it over the place, and particularly around the edges, to remove the border that would still present a stain.

Materials of Which Fish Are Composed.

"Personal"

[Affixed label with image of Argentine flag and words "Argentine Republic."]

Fred S. Calhoun

2nd Lieut. 14th Infantry

U.S. Army.

[Affixed newspaper clipping titled "To Remove Grease Spots from Paper"]

Headquarters Dept. of Dakota
Saint Paul, Minn., Dec'r 29" 1874

Special Orders
No. 283 } Extract.

I In accordance with instructions from the War Department a Board of Officers to Consist of

Lieut Col W. P. Carlin 17th Infty
Surgeon J. F. Weeds U.S.A.
Captain G. B. Dandy A.Q.M. U.S.A
Captain G. W. Yates 7th Cavalry

is hereby appointed to meet at Fort Abraham Lincoln, D.T., on Wednesday the 13th day of Jan'y, 1875, or as soon thereafter as practicable for the examination of Mr F. A. Calhoun, and such other candidates, who have been selected for the appointment of 2d Lieutenant in the army of the United States, as may be properly authorized to appear before it.

The duties of the Medical Officer will be confined to the Medical examination.

The junior member will act as Recorder

× × × × × × ×

By Command of Brig' Gen'l Terry
"Signed" O. D. Greene
A. A. General

O. D. Greene

Mar 19th 1875.
Received Commission & ordered to Newport Barracks Ky —

Fred S. Calhoun
U.S.A.

Headquarters Dept of Dakota
Saint Paul, Minn, Decbr 29th 1874

Special Orders
No 283 -   Extract.
          I   In accordance with instructions from the War Department a Board of Officers to Consist of

   Lieut. Col. W.P. Carlin, 17th Inftry
   Surgeon JF. Weeds U.S.A.
   Captain G.B. Dandy A.Q.M. U.S.A.
   Captain G.W. Yates 7th Cavalry

is hereby appointed to meet at Fort Abraham Lincoln, D.T., on Wednesday the 13th day of Jany, 1875, or as soon thereafter as practicable for the examination of Mr. F.S. Calhoun, and such other candidates who have been selected for the appointment of 2d Lieutenant in the Army of the United States, as may be properly authorized to appear before it.

The duties of the Medical Officer will be confined to the Medical examination. The junior member will act as Recorder

X   X   X   X   X   X

By Command of Brig. Genl Terry
"Signed" O.D. Greene
A.A. General

O.D. Greene

Mar 19th 1875.
Received Commission & ordered to Newport Barracks Ky -
Fred S. Calhoun
U.S.A.

> Headquarters Dept of Dakota
> Saint Paul, Minn., March 23," 1875.
>
> Special Orders }
> No 46 } Extract.
>
> I In accordance with authority from the Lieut General Comdg the Military Division of the Missouri, 2d Lieutenant F. P. Calhoun, 14" Infantry, (recently appointed) will report to the Commanding Officer Fort Abraham Lincoln, D.T., for temporary duty at that post until such time as the opening of the Northern Pacific Railroad shall enable him to proceed to join his company.
>
> × × × × × × ×
>
> By Command of Brig Genl Terry
> "sgd" O. D. Greene
>   a.a. General
>
> Official
> "sgd" R. P. Hughes
>   Captain 3" Infty
>   A.D.C.
>
> Recd April 11" 75.

Headquarters Dept. of Dakota
Saint Paul, Minn., March 23rd, 1875
Special Orders
No 46 -　　　　Extract
　　I. In accordance with authority from the Lieut. General Com'd of the Military Division of the Missouri, 2nd Lieutenant F. P. Calhoun, 14th Infantry, (recently appointed) will report to the Commanding Officer Fort Abraham Lincoln, D.T., for temporary duty at that post until such time as the Opening of the Northern Pacific Railroad shall enable him to proceed to join his Company.
X　　X　X　X　　X　X　X
By command of Brig Gen'l Terry
　"Signed"　　O.D. Greene,
　　　　　　A.A. General
Official
　"Signed R.P. Hughes
　　　Captain 3rd Inftry
　　　　A.D.C.

Rec'd April 11th 75.

Hdqrs Fort A Lincoln D.T.
April 12" 1875.

Special Orders) Ext.
No 56)

I. 2nd Lieut F.S. Calhoun 14th Infty, having reported at these Headqrs in compliance with S.O. No 46 Hdqrs Dept of Dakota, will report to the Comdg Officer "B" Co 6" Infty for duty.

By order of Bvt Maj Genl Custer

"Sgd" James Calhoun
1" Lieut 7th Cavaly
Post Adjutant.

War Department. A. G. O.
Washington, April 3, 1875.

Special Orders)
No 56) Extract.

IIII. So much of S.O. No. 46, March 23, 1875, from Hd. Qrs Dept of Dakota, as directed F.P. (F.S.) Calhoun as 2nd Lieutenant, 14th Infantry, to report to the Commanding Officer Fort Abraham Lincoln, Dakota Terry, for duty, is hereby revoked; the said F.P. (F.S.) Calhoun, at the date of said order, not being in the United States Service nor in the receipt of his Commission thus to place him in the service of the United States.

× × × × × × × × ×

By order of the Secretary of War

Official:
"Sgd" L.H. Pelouze
a.a. Genrl.

E. D. Townsend,
Adjutant General.

Received at Hd qrs
April 19" 1875

Hdqrs Fort A Lincoln D.T.
April 12th 1875.

Special Orders
No 56 -   Ext.

I. 2nd Lieut. F.S. Calhoun 14th Inftry, having reported at these Headqrs in compliance with S.O. No 46 Hdqrs Dept of Dakota, will report to the Comdg Officer "B" Co 6" Inftry for duty.
By order of
    Bvt Maj Genl Custer
"Sgd" James Calhoun
    1st Lieut 7th Cavalry
    Post Adjutant

[New Entry:] War Department. A.G.O.
    Washington, April 3, 1875

Special Orders
No 56 - Extract

IIII So much of S.O. No. 46, March 23, 1875, from Hd.Qrs Dept of Dakota, as directed F.P. (F.S.) Calhoun as 2nd Lieutenant, 14th Infantry, to report to the Commanding Officer Fort Abraham Lincoln, Dakota Terr, for duty, is hereby revoked, the said F.P. (F.S.) Calhoun, at the date of said order, not being in the United States Service nor in the receipt of his commission thus to place him in the service of the United States.

X   X   X   X   X   X X X X
    By order of the Secretary of War
Official: E.D. Townsend,
"Sgd" L.H. Pelouze   Adjutant General
A.A. General [In light red handwriting in the bottom corner:]
    Received at Ft. A. Lincoln April 17th 1875

Head qrs Fort A Lincoln D.T.
April 18" 1875.

Special Orders) Extract.
No 60 }

II Special Orders No 56 par
I Current Series from these Headquarters
is hereby revoked and 2nd Lieutenant
F.S. Calhoun 14" Infantry will Comply
with instructions received by him from
the War Department.

By Order of Bvt Maj Genl Custer

"Sgd" James Calhoun
1st Lieut 7th Cavalry
Bvt Adjutant.

Left Fort Lincoln D.T.
April 26" 1875 on first train N.P.R.R.
& in Compliance with this Order
Reported at Newport
Barracks Ky. May 19" 1875
(1875)

Hd.-Qrs. Depot Genl Recty Service.
Newport Barracks Ky
May 25" 1875.

Special Orders) Ext.
No 77 }

x x x x x x x x x

III The following named Commissioned
Officers are hereby assigned, respectively,
to the Companies set opposite their names,
and will report to the Company Commanders
for duty:

1st Lieut J.A. Haughey, 21st Infty, to Co "C"
2nd Lieut F.S. Calhoun, 14" Infty, to Co "E"

By Order of Maj Edwin C. Mason

"Sgd" L. M. Morris
1st Lieut 20" Infty
Depot Adjutant.

Headq'rs Fort A Lincoln D.T.
April 18th 1875

Special Orders
 No 60 -   Extract
        II Special Orders No 56 para
I Current Series from these Headquarters
is hereby revoked and 2nd Lieutenant
F.S. Calhoun 14th Infantry will comply
with instructions received by him from
the War Department.
By order of Bvt Maj Genl Custer
 "Sgd" James Calhoun
1st Lieut 7th Cavalry
Post Adjutant
[In red pencil:] Left Fort Lincoln D.T. April 26th 1875 on first [illegible] NPRR. in compliance with this order Reported at Newport Barracks Ky. May 19th 1875

Hd.Qrs. Depot Genl Recty Service.
Newport Barracks Ky
May 25th 1875

 Special Orders
 No 77 - Ext
X   X   X   X   X   X   X   X   X
III The following named Commissioned
Officers are hereby assigned, respectively,
to the Companies set opposite their names,
and will report to the company Commanders
for Duty:
1st Lieut J.A. Haughey, 21st Inftry, to Co "C"
2nd Lieut F.S. Calhoun, 14th Inftry, to Co "E"

By order of Maj Edwin C. Mason

"Sgd" L.M. Morris 1st Lieut 20th Inftry
Depot Adjutant.

Hd. Qrs. Dept Genl. Rectg Service
Newport Barracks, Ky. May 26" 1875.

Special Orders }
No 78 } I. A Board of Survey will convene at this Depot at 12 O'clock M., today, or as soon thereafter as practicable, to examine into and fix the responsibility of a deficiency of a number of Shoes found in the Quartermasters Dept, and for which 1st Lieut F.F. Riley, 21st Infty A.A.Q.M. is responsible.

Detail for the Board
1st Lieut J. M. Thompson, 21st Infantry
1st Lieut G. A. Young, 4th Infantry
1st Lieut F.S. Calhoun, 14th "

By order of Maj: Edwin C. Mason

Sgd L. M. Morris
1st Lieut 10th Infty
Depot Adjutant

War Department A.G.O.
Washington, May 31" 1875.

Special Orders }
No 106 } Extract
5. A General Court Martial is appointed to meet at Newport Barracks, Kentucky, on the 8th day of June, 1875, or as soon thereafter as practicable, for the trial of Private David Van Pelt, Company "F" 16th Infty, and such other prisoners as may be brought before it.

Detail for the Court.
Maj: E.C. Mason, 21st Infty
1st Lieut J.A. Haughey 21" "
2nd " H.F. Cunningham 8" "

Over

Hd.Qrs. Depot Genl. Rectg Service
Newport Barracks, Ky, May 26th 1875
Special Orders
No 78   I A Board of Survey will convene
at this Depot of 12 O'Clock inst. [*intante mense*, this month],
today,
or as soon thereafter as practicable, to
examine into and fix the responsibility[?]
of a deficiency of a number of shoes
found in the Quartermasters Dept, and
for which 2st Lieut J. F. Riley, 21st Inftry
A.A.S.M. is responsible.
    Detail for the Board
1st Lieut. J.M. Thompson, 24th Infantry
2nd Lieut G.S. Young, 7th Infantry
2nd Lieut F.S. Calhoun, 14th Infantry
    By order of Maj Edwin C. Mason

Sgd L.M. Morris
1st Lieut 20th Inftry
Depot Adjutant
    War Department A.G.O
    Washington May 31, 1875
Special Order
No 106 - Extract
        5. A General Court Martial is
appointed to meet at Newport Barracks,
Kentucky on the 8th day of June, 1875, or
as soon thereafter as practicable, for the
trial of Private David Van Pelt, Company
"F," 16th Inftry, and such other prisoners
as may be brought before it.
    Detail for the Court.
 Maj E.C. Mason, 21st Inftry
 1st Lieut J.A. Haughey 21st " "
 2nd Lieut N.F. Cunningham 8th " "

    Over [List continued on page 8]

Custer's Other Brother-in-Law

8

Continued.

2nd Lieut G.S. Young, 7th Infty
2nd Lieut F.S. Calhoun, 14" "
1st Lieut J.M. Thompson, 4th " Infty, Judge Advocate of the Court.

No other officers than those named can be assembled without manifest injury to the Service.

The Court is authorized to sit without regard to hours.

× × × × ×

By order of the Secretary of War
E.D. Townsend
Adjutant General

Official:
Sgd L.H. Pelouze
A.A.G.

War Department A.G.O.
Washington, June 7th 1875.

Special Orders) Extract.
No 113)

IIII Surgeon Ebenezer Swift and 1st Lieut Patrick Cusack, 9th Cavalry, are detailed as members of the General Court Martial appointed by Special Orders No 106 May 31, 1875, from this Office, to meet at Newport Barracks, Kentucky, on the 8th instant, Vice 2d Lieutenants G.S. Young 7th Infantry and F.S. Calhoun, 14th Infty, hereby relieved.

By order of the Secretary of War
E.D. Townsend
Adjutant General.

Official:
Sgd L.H. Pelouze
A.A.G.

Continued.
  2nd Lieut G.S. Young, 7th Inftry
  2nd Lieut FS. Calhoun 14th Inftry
1st Lieut J.M. Thompson 24th Inftry, Judge Advocate of the Court.
  No other officers than those named can be assembled without manifest injury to the Service.
The Court is authorized to sit without regard to hours.
  X   X   X   X   X
  By order of the Secretary of War
        E.D. Townsend
        Adjutant General
Official:
  "Sgd" LH Pelouze
      A.A.G.

            War Department A.G.A.
            Washington June 7th 1875
  Special Orders
No 113 -   Extract
  IIII surgeon Ebenezer Swift and 1st Lieut Patrick Cusack, 9th Cavalry, are detailed as members of the General Court Martial appointed by Special Orders No 106 May 31, 1875, from this Office, to meet at Newport Barracks, Kentucky, on the 8th inst. Vice 2d Lieutenant GS. Young 7th Infantry and FS. Calhoun, 14th Inftry, hereby relieved.
            By order of the Secretary of War
            E.D. Townsend Adjutant General.
Official
"sgd" L.H. Pelouze
      A.A.G.

Hd. Qrs. Depot Genl. Recty. Service
Newport Barracks, Ky., June 9th 1875.

Special Orders {
No 86 }
I A Garrison Court Martial will assemble at this Depot at 10 o'clock A.M., today, or as soon thereafter as practicable, for the trial of such prisoners as may be properly brought before it.

Detail for the Court.
1st Lieut. J. A. Haughey, 21st Infty
2nd Lieut. Jas. Young, 7th Infty
2nd Lt. F. S. Calhoun, 14th "

By order of Maj. Edwin C. Mason
Signed L. M. Morris
1st Lieut. 10th Infty
Depot Adjutant.

Hd. Qrs. Genl. Recty. Service
Newport Barracks, Ky. June 12th 1875.

Special Orders {
No 89 } Extract
II In compliance with S.O. No 97, Current Series, Hd. Qrs. G.R.S., New York City, 2d Lieut. F. S. Calhoun, 14th Infty, will take charge of two (2) Colored Band Musicians, and fifty five (55) Colored Recruits, and proceed with them at once, to St Louis Barracks, Mo., where, upon arrival, they will be turned over to the Commanding Officer of that Post for assignment to the 25th Infantry.

III Mrs. Nevil T. Henderson is hereby appointed Laundress to the above detachment.

Upon completion of the above duty, Lieut. Calhoun will return without

Over

Hd.Qrs Depot Genl Recty Service
Newport Barracks, Ky., June 7th 1875
[illegible] Orders
No 86   I A Garrison Court Martial will assemble at this Depot at 10 O' clock AM, today, or as soon thereafter as practicable, for the trial of such prisoners as may be properly brought before it.
  Detail for the Court
1st Lieut J.A. Haughey, 21st Inftry
2nd Lieut G.S. Young, 7th Inftry
2nd Lt FS. Calhoun 14th Inftry
 By order of Maj Edwin C. Mason
   signd LM Morris
    1st Lieut 20th Inftry
      Depot Adjutant

Hd Qrs Genl Recty Service
Newport Barracks Ky June 12th 1875
Special Orders
No 89 -   Extract
 II   In compliance with S.O No 97, current Series, Hd.Qrs. GR.S., New York City, 2d Lieut F.S. Calhoun, 14th Inftry, will take charge of two (2) Colored Band Musicians, and fifty five (55) colored Recruits, and proceed with them at once, to St. Louis Barracks, Mo., where, upon arrival, they will be turned over to the Commanding Officer of that Post for assignment to the 25th Infantry.
III   Mrs. Nevil T. Henderson is hereby appointed Laundress to the above detachment. Upon completion of the above duty Lieut Calhoun will return without

Over

10
Continued

delay to this Depôt.
The detachment will be furnished
with one (1) days cooked rations.
The Quartermaster Department will
furnish the necessary transportation.
By order of Maj. Edwin C. Mason

"Sgd" L. M. Morris
1st Lieut. 20th Infantry
Depot Adjutant

Head Quarters Genl Rectg Service
United States Army
New York City, June 16" 1875.

Special Orders } Extract
No. 103

II 2nd Lieut. F. S. Calhoun, 14th Infty,
is hereby relieved from duty at Newport
Barracks, Ky., and will proceed
without delay to Principal Depot
Governors Island, N.Y.H. reporting upon
his arrival there, to the Commanding
Officer of the Depot, thence to proceed
to join the 14th U.S. Infantry, with
detachment of recruits under orders
for that Regiment, in the Department
of the Platte.
By Command of Colonel King
"Signed" Horace Neide
1st Lieut. 4th Infantry
A.A.A. General.

Left Newport Barracks June 23" in Compliance with
above order - Reported at Governors Island N.Y.H.
June 24" 1875

Continued
delay to this Depot.
The detachment will be furnished
with one (1) days Cooked rations.
The Quartermasters Department will
furnish the necessary transportation.
By order of Maj Edwin C. Mason
    "Sgn" L.M. Morris
        1st Lieut 20th Infantry
            Depot Adjutant

    Headquarters Genl Recty Service
        United States Army
    New York City, June 19th 1875
Special Orders
No 103 - Extract
    II 2nd Lieut F.S. Calhoun 14th Inftry,
    is hereby relieved from duty at Newport
    Barracks, Ky., and will proceed
    without delay to Principal Depot
    Governors Island, NY.N. reporting upon
    his arrival there, to the Commanding
    Officer of the Depot, thence to proceed
    to join the 14th U.S. Infantry, with
    detachment of recruits under orders
    for that Regiment, in the Department
    of the Platte.
        By Command of Colonel King
"Signed"   Horace Neide
        1 Lieut 5th Infantry
        A.A.A. General.

[In red pencil:] Left Newport Barracks June 23rd in compliance with above order - Reported at Governors Island N.Y.N. June 24th 1875

Hd. Qrs. Depot Genl Rectg Service
Newport Barracks, Ky., June 21" 1875

Special Orders)
No 95   } I To enable him to comply
with the requirements of Par. 2, S.O. No
103. C.S. Hd. Qrs. G.R.S. New York City
2d Lieut F.S. Calhoun, 14" Infantry, is
hereby relieved from duty at this Depot.
By order of Lieut J.M. Thompson

"Sgd"   N.F. Cunningham
        1st Lieut 8" Infantry
        Acting Adjutant.

Headquarters, Principal Depot G.R.S.
Fort Columbus, New York Harbor
June 29" 1875.

Special Orders)
No 161   }   Extract
XIII In Compliance with instructions
from Headquarters General Recruiting
Service, a detachment of Recruits,
Artificers and Musicians will leave
this Depot on Friday, the 2d of July 1875,
for Omaha, Nebraska.

170 Recruits for 14th Infantry
  1 Trumpeter Co. B. 14" Infantry
  1   "   " F. "   "
  1   "   " N. "   "
  1 Fifer   " C. "   "
  1 Band Musician "   "   "
  1 Trumpeter Co. B. 9" Infantry
  1   "   " N. "   "
  1   "   " D. 4th   "
  2   "   " E. "   "
                                Over.

Hd.Qrs. Depot Genl Rectg Service
Newport Barracks, Ky., June 21st 1875

Special Orders
No 95   I To enable him to comply with the requirements of Par. 2, S.O. No 103, C.S. Hd. Qrs. G.R.S. New York City 2d Lieut FS. Calhoun, 14th Infantry, is hereby relieved from duty at this Depot.

By order of Lieut J.M. Thompson
"Sgd" NF. Cunningham
2nd Lieut 8th Infantry
Acting Adjutant.

Headquarters, Principal Depot G.R.S.
Fort Columbus, New York Harbor
June 29th 1875.

Special Orders
No 161 -   Extract

XIII In compliance with instructions from Headquarters General Recruiting Service, a detachment of Recruits, Artificers[?] and Musicians will leave this Depot on Friday, the 2d of July 1875, for Omaha, Nebraska.

170 Recruits for 14th Infantry
1 Trumpeter Co. B 14th Infantry
1 Trumpter Co. F. 14th Infantry
1 Trum[eter Co. H. 14th Infantry
1 Fifer Co. C. 14th Infantry
1 Band Musician 14th Infantry
1 Trumpeter Co B. 9th Infantry
1 Trumpeter Co H. 9th Infantry
1 Trumpeter Co D. 4th Infantry
2 Trumpeters Co E. 4th Infantry

Over

12
Continued

1	Tailor	Co	E.	4th	Infantry
1	Carpenter	"	"	"	"
1	"	"	D.	23rd	"
2	"	"	N.	"	"
1	Shoemaker	"	"	"	"
1	"	"	F.	"	"

Captain M. J. Fitzgerald, 9th Infantry is assigned to the Command of the detachment. He will be assisted in his duties by First Lieut. T. F. Riley, 21st Infantry, and 2d Lieut F. S. Calhoun, 14th Infantry.

Act. Asst. Surgeon A. F. Steigers, U. S. A. is assigned as Medical Officer to accompany the detachment to Omaha Neb. He will report to the Commanding Officer on the morning of Friday, the 2d proximo.

The detachment will be supplied with (4) days Cooked rations.

Pursuant to instructions from the A. G. O. dated October 26th. 1869, the Acting Commissary of Subsistence of the Depot will turn over to Captain Fitzgerald Two Hundred ($200.) dollars subsistence funds, to be expended in the purchase of two (2) quarts of liquid Coffee to each man per day. The funds & Coffee to be accounted for to the Subsistence Department in the usual manner.

A. Model Muster Roll will be furnished to the Officer in Command for his guidance. This Roll exhibits all the Changes incident to the transfer of recruits to regiments; it together

Continued
   1 Tailor co E. 4th Infantry
   1 Carpenter Co E. 4th Infantry
   1 Carpenter Co D. 23rd Infantry
   2 Carpenters Co H. 23rd Infantry
   1 Shoemaker Co H. 23rd Infantry
   1 Shoremaker Co F. 23rd Infantry
Captain M J Fitzgerald, 9th Infantry
is assigned to the Command of the
detachment. He will be assisted
in his duties by First Lieut. J.F. Riley,
21st Infantry, and 2d Lieut F.S. Calhoun,
14th Infantry.
  Act. Asst. Surgeon AF. Steigers, U.S.A.
is assigned as Medical Officer to
accompany the detachment to Omaha
Neb. He will report to the Commanding
Officer on the morning of Friday,
the 2d proximo. [the month following the present.]
The detachment will be supplied
with (4) days cooked rations.
Pursuant to instructions from the
A.G.O. dated October 26th, 1869, the
Acting Commissary of the subsistence of
the Depot will turn over to Captain
Fitzgerald Two Hundred ($200) dollars
subsistence funds, to be expended
in the purchase of two (2) quarts
of liquid coffee to each man per day.
The funds & Coffee to be accounted
for to the Subsistence Department
in the usual manner.
A. Model Muster Roll will be furnished
to the Officer in Command for his
guidance. This Roll exhibits all
the Changes incident to the transfer
 of recruits to requirements; it together

Continued

with the regulations governing the Recruiting Service, will be strictly adhered to in all that relates to reports & returns.

Upon arrival at Omaha, Neb., the Medical Officer will be relieved from duty with the detachment, and Captain Fitzgerald will report his command to the Commanding General of the Department of the Platte.

The Quartermasters Department will furnish the necessary transportation.

XIV Upon being relieved from duty with the detachment at Omaha, Neb., A. A. Surgeon Steigers, U.S.A. will return to his proper station, Fort Columbus, New York Harbor.

By Order of Maj. R. E. A. Crofton
"Signed" C. S. Roberts
1st Lieut 14th Infty
Post Adjutant

Left N.Y.H. July 2nd 1875
Arrived at Camp Douglas U.T.
July 8th 1875

Headquarters Dept of the Platte
Omaha, Neb., July 5th 1875.

Special Orders No 75 | Extract

III The Quartermasters Department will furnish transportation from this City to Camp Douglas, U.T. for Captain Michael J Fitzgerald, 9th Infantry, First Lieut. F. R. Riley, 21st Infty, Second Lieut F. S. Calhoun, 14th Infty, and and One hundred and Seventy six recruits for 14th Infantry; to Fort D. A. Russell for two recruits 9th Infantry, five recruits 14th Infty

Over

Continued
with the regulations governing the
Recruiting Service, will be strictly
adhered to in all that relates to
reports & returns.
Upon arrival at Omaha, Neb., the Medical
Officer will be relieved from duty with
the detachment, and Captain Fitzgerald
will report his command to the
Commanding General of the Department
of the Platte.
The Quartermasters Department will
furnish the necessary transportation.
XIV Upon being relieved from duty with
the detachment of Omaha, Neb.,
A.A. Surgeon Steigers, U.S.A. will return
to his proper station, Fort Columbus,
New York Harbor
By Order of Maj R.E.A. Crofton
" signed" C.S. Roberts 1st Lieut.
    17th Inftry
     Post Adjutant
[In red pencil:] Left NYN. July 2nd 1875 arrived at Camp Douglas UT July 8th 1875

   Headquarters Dept of the Platte
    Omaha, Neb, July 5th 1875
Special Orders
No 75 - Extract
III The Quartermasters Department
will furnish transportation from this
City to Camp Douglas, UT. after Captain
Michael J Fitzgerald, 9th Infantry,
First Lieut. T.F. Riley, 21st Inftry, Second
Lieut F.S. Calhoun, 14th Inftry, and
and one hundred and seventy six recruits
for 14th Infantry; to Fort D.A. Russell for two recruits
9th Infantry, five recruits 4th Infantry
         Over

14

Continued

and four recruits 23rd Infantry; and to Fort Fred Steele for one recruit 23rd Infantry.
By Command of Brigadier General Crook
"Signed" Geo. D. Ruggles
A.A. General

Headquarters 14th U.S. Infantry.
Camp Douglas U.T. July 9th 1875.

Special Orders } Extract
No 50

II 2d Lieut F.S. Calhoun having reported at these Headquarters by virtue of G.O. No 47, Dated A.G.O April 1st /75, is hereby assigned to Company "F" 14" Infantry, to the Commanding Officer of which Company he will report without unnecessary delay.

By Order of Major M. Bryant.
"Syd" 10th W. McCammon
1st Lt & Adjt 14" Infty

Hd. Qrs. Camp Douglas U.T.
July 14" 1875.

Special Order } A Garrison Court Martial to
No 95   consist of
  Captain Kennington }
  1st Lieut Quentin   } 14" Infty
  2nd Lieut Calhoun   }
will convene at this Post at 10. o'clock AM. tomorrow the 15" inst: or as soon thereafter as practicable for the trial

Continued
and four recruits 23rd Infantry; and
to Fort Fred Steele for one recruit
23rd Infantry.
  By Command of Brigadier General Crook
      "signed" Geo D. Ruggles
          A.A. General

      Headquarters 14th U.S. Infantry
      Camp Douglas U.T. July 9th 1875
Special Orders
No 50 -   Extract
II 2nd Lieut F.S. Calhoun having
reported at these Headquarters by
virtue of G.O. No 47, Dated A.G.O.
April 1st 75, is hereby assigned to
Company "F" 14th Infantry, to the
Commanding Officer of which
Company he will report without
unnecessary delay.
        By order of Major M. Bryant.
  "Sgd" Wm W McCammmon
      1st Lt & Adjt 14th Inftry

        Hd. Qrs. Camp Douglas U.T.
          July 14th 1875

Special Orders
No 95   A Garrison Court Martial to
Consist of Captain Kennington
1st Lieut Quentin   14th Inftry
2nd Lieut Calhoun
 will convene at this Post at 10 o'clock
AM tomorrow the 15th inst. or as soon
thereafter as practicable for the trial

Continued

of such persons as may be properly brought before it.

By order of Major Bryant

Sgd. Howard M Cannon
1st Lt & Adjt 14" Infy & Post

Headquarters Camp Douglas U.T.
July 22nd 1875.

Special Orders No 101 } A Board of Survey to consist of Captain D. W. Burke, 1st Lieut J. E. Quentin, 2nd Lieut R. S. Calhoun } 14" Infantry will convene at this Post at 10-30 O,C AM. the 23" inst. to determine in regard to a deficiency between the marked and actual weight of one barrel Cut Loaf Sugar — for which 2d Lieut John Murphy 14" Infy A.C.S is responsible and to fix the responsibility thereon. Also to report upon the quality & condition of a lot of Java Coffee received from Chief C.S. Dept of the Platte on the 9th inst.

By order of Maj Bryant

Sgd. Howard M Cannon
1st Lt adj 14" Infty & Post

Continued
of such persons as may be properly
brought before it.
> By order of Major Bryant
> sgd" Wm W McCammmon
> 1st Lt & Adj 14th Inftry & Post

> Headquarters Camp Douglas U.T.
> July 22nd 1875

Special Orders
No 101   A Board of Survey to Consist of
Captain D.W. Burke
1st Lieut J.E. Quentin   14th Infantry
2nd Lieut F.S. Calhoun,
will convene at this Post at 10:30 O,C
AM the 23rd wish to determine in regard
to a deficiency between the marked
and actual weight of one barrel
cut loaf Sugar - for which 2d Lieut
John Murphy 14th Infantry A.C.S. is responsible
and to fix the responsibility thereon.
Also to report upon the quality &
condition of a lot of Java Coffee
received from Chief CS. Dept of the
Platte on the 7th inst. .
> By order of Maj Bryant
> "sgd" Wm W McCammon
> 1st Lt Adj 14th Inftry & Post.

16

Headquarters Camp Douglas U.T.
August 19" 1875.

Special Orders
No 120 } II A Garrison Court Martial
to Consist of
Captain James Kennington
1st Lieut R. P. Warren } 14" Infty
2nd Lieut F. S. Calhoun
will convene at this Post at 10 O'C
A.M. tomorrow the 20th inst: or as soon
thereafter as practicable, for the trial
of such persons, as may be properly
brought before it.

By order of Capt Burke
"Sig" W W M°Cammon
1st Lt & Adjt 14" Infty & Post.

Headquarters Camp Douglas U.T.
August 22nd 1875.

Special Orders
No 122 } III Extract.
2nd Lieut F. S. Calhoun is hereby
detached from Co F 14" Infty and
will report to Capt D. W. Burke 14"
Infty ordered to Corinne Utah.

By order of Colonel Smith
"Sig" W. W. M°Cammon
1st Lt & Adjt 14" Infty & Post.

Arrived at Corinne Utah August 23" 1875.
Returned to Camp Douglas Sept 23" 1875.
One Month absent.

Headquarters Camp Douglas U.T.
August 19th 1875

Special Orders
No 120   II A Garrison Court Martial
to Consist of
Captain James Kensington
1st Lieut R.P. Warren   14th Inftry
2nd Lieut FS. Calhoun
will convene at this Post at 10 O,C
AM tomorrow the 20th inst. or as soon
thereafter as practicable, for the trial
of such persons, as may be properly
brought before it.
   By order of Capt Burke
   "Sgd" Wm W McCammon
   1st Lt & Adjt 14th Inftry & Post.

Headquarters Camp Douglas U.T.
August 22nd 1875

Special Orders
No 122 - Extrct III
2nd Lieut F.S. Calhoun is hereby
detached from Co F 14th Inftry and
will report to Capt. D.W. Burke 14th
Inftry ordered to Corinne, Utah.
   By order of Colonel Smith
   "Sgd" W.W. McCammon
   1st Lt & Adjt 14th Inf & Post

[In red pencil:] Arrived at Corinne Utah August 23rd 1875.
Returned to Camp Douglas Sept 23rd 1875.
One Month absent.

Headquarters Camp Douglas U.T.
October 1" 1875.

Special Orders } A Board of Survey to consist of
No 144
Captain G. S. Carpenter
1" Lieut R. P. Warren } 14" Infty
2" Lieut F. S. Calhoun
is hereby ordered at this Post at
10 o'clock A.M. tomorrow the 2nd inst
or as soon thereafter as practicable for
the purpose of Examining into and
reporting upon the quantity and quality
of a certain amount of Coal, to be
delivered by the U.C.R.R.
By order of Maj Bryant
"Sgd" Wm W. McCammon
1" Lieut & Adjt 14" Infty & Post.

Headquarters Dept of the Platte
Omaha Neb. October 29" 1875.

Special Orders } Extract
No 119
I A general Court Martial is
hereby appointed to meet at Camp
Douglas, U.T. at 10 o'clock A.M.
on Wednesday, the 3rd day of
November, 1875, or as soon thereafter
as practicable, for the trial of such
persons as may be properly brought
before it.
Detail for the Court.
Major Montgomery Bryant   14" Infty
Captain Gilbert S. Carpenter   "
   "   Daniel W. Burke   "
   "   James Kennington   "
   "   Thomas F. Tobey   "
                              Over.

Headquarters Camp Douglas U.T.
October 1st 1875

Special Orders
No 144   A Board of Survey to Consist of
Captain G.S. Carpenter
1st Lieut R.P. Warren   14th Infantry
2nd Lieut. F.S. Calhoun,
is hereby ordered at this Post at
10 Oclock AM tomorrow the 2nd inst.
or as soon thereafter as practicable for
the purpose of Examining into and
reporting upon the quantity and quality
of a certain amount of Coal, to be
delivered by the U.C.RR.
    By order of Maj Bryant
    "Sgd" Wm W McCammon
        2st Lieut & Adjt 14th Inftry & Post.

Headquarters Dept of the Platte
Omaha Neb October 27th 1875

Special Orders
No 117 -   Extract
I A general Court Martial is
hereby appointed to meet at Camp
Douglas, U.T. at 10 O'Clock AM
on Wednesday, the 3rd day of
November, 1875, or as soon thereafter
as practicable, for the trial of such
persons as may be properly brought
before it.
detail for the Court
Major Montgomery Bryant 14th Intry
Captain Gilbert S. Carpenter 14th Inftry
Captain Daniel W Burke 14th Inftry
Captain James Kensington 14th Inftry
Captain Thomas F. Tobey 14th Inftry   Over.

Continued.

2d Lieut Charles A. Johnson 14" Infty
 "   Charles F. Lloyd   "
 "   Fredk S. Calhoun   "
 "   Jos H Gustin   "
Second Lieut Richard T. Yeatman
14" Infty Judge Advocate.
No other officers than those named
can be assembled without manifest
injury to the service.
    By Command of Brig Gen Crook
"Sgd" Geo D Ruggles
         A. A. General

                Headquarters Camp Douglas U.T.
                     November 4" 1875.
Special Orders
No 161   A Board of Survey to consist of
Captain Paul W. Burke
1st Lieut R.P. Warren   } 14" Infty
2d Lieut F.S. Calhoun
is hereby ordered at this Post
at 10 O'clock tomorrow the 5" inst:
or as soon thereafter as practicable
for the purpose of determining in
regards to deficiencies found to
exist during the past month in
certain original packages of subsistence
stores opened for issue to Troops
and for which 2d Lieut John
Murphy 14 Infty A.C.S. is responsible.
         By order of Colonel Smith
"Sgd" Wm W McCammon
      1st Lieut & Adjt 14" Infty & Post

Continued
2d Lieut Charles A Johnson 14th Inftry
2d Lieut Charles F. Lloyd 14th Inftry
2d Lieut Fredk S. Calhoun 14th Inftry
2d Lieut Joseph H Gustin 14th Inftry
Second Lieut Richard T Yeatman
14th Inftry Judge Advocate.
No other officers than these named
Can be assembled without manifest
injury to the Service.
  By command of Brig Gen Crook
     "Sgd" Geo D Ruggles
      A.A. General

        Headquarters Camp Douglas U.T.
          November 4th 1875
Special Orders
No 161   A Board of Survey to Consist of
Captain Paul W. Burke
1st Lieut RP. Warren   14th Inftry
2nd Lieut. FS. Calhoun
is hereby ordered at this Post
at 10 O'clock tomorrow the 5th inst.:
or as soon thereafter as practicable
for the purpose of determining in regards to
deficiencies found to
exist during the past month in
Certain original packages of [illegible]
Shirts offered for issue to Troops
and for which 2nd Lieut John
Murphy 14th Infrty A.C.S. is responsible.

     By order of Colonel Smith
   "Sgd" Wm W McCammon
     1st Lieut & Adjt 14th Inftry & Post.

Headquarters Camp Douglas U.T.
September 24th 1875.

Special Orders
No 141   &c
II 2nd Lieut F.S. Calhoun 14th Infty is hereby relieved from duty with Company "C" 14th Infty and will report to his Company Cmdr for duty.
By order of Maj Bryant
"sgd"   Wm W. McCammon
1st Lieut & Adjt 14th Infty & Post.

NOTE This order properly belongs on Page 14" but was omitted — Received verbal order same date to do duty with C. Co.
Fred S. Calhoun, U.S.A.

Headquarters Camp Douglas U.T.
December 12" 1875.

Special Orders
No 182   A Board of Survey to consist of
Captain D.W. Burke
1st Lieut R.P. Warren } 14th Infty
2nd Lieut F.S. Calhoun
is hereby ordered at this Post at 10 O'clock tomorrow the 13" inst or as soon thereafter as practicable for the purpose of determining in regard to deficiencies found to exist during the past month in certain original packages of subsistence stores, opened for issue to troops and for which 2nd Lieut John Murphy 14" Infty A.C.S. is responsible.
By order of Colonel Smith
"sgd"   Wm W. McCammon
1st Lt & Adjt 14" Infty & Post

Headquarters Camp Douglas UT.
September 24th 1875

Special Orders
No 141 -   Ex
II 2nd Lieut F.S. Calhoun 14th Inftry is
hereby relieved from duty with Company "C"
14th Inftry and will report to his Company Comdr
for duty
                By order of Maj Bryant
"Sgd"   Wm W McCammon
                1st Lieut & Adjt 14th Inftry & Post.
 [In red pencil:] <u>Note</u> This order properly belongs on page 17th
but was omitted - Received verbal order same
date to do duty with C. Co.
                      FS. Calhoun U.S.A.

Headquarters Camp Douglas U.T.
December 12 1875

Special Orders
No 182   A Board of Survey to Consist of
Captain DW Burke
1st Lieut R.P. Warren   14th Inftry
2nd Lieut F.S. Calhoun,
is hereby ordered at this Post at 10'OClock
tomorrow the 13th inst. or as soon thereafter
as practicable for the purpose of determining
in regard to deficiencies found to exist
during the past month in certain [illegible]
packages of Subsistence Stores, offered for
issue to troops and for which 2nd Lieut John
Murphy 14th Inftry A.C.S. is responsible.
                By order of Colonel Smith
       "Sgd"   Wm W McCammon
            1st Lt & Adjt 14th Inftry & Post

Headquarters Camp Douglas U.T.
January 19th 1876

Special Orders }
No 8

A Garrison Court Martial is hereby ordered to convene at this Post at 10 o'clock the 20' inst. or as soon thereafter as practicable for the trial of such Prisoners as may be properly brought before it.

Detail for the Court
Capt Thos F. Tobey   }
1st Lieut I. E. Quentin   } 14" Infty
2nd Lieut F. S. Calhoun   }

By Order of Col Smith
(sgd) Wm. C. McCammon
1st Lt & Adjt. 14" Infty & Post.

Headquarters Camp Douglas U.T.
February 1st 1876.

Special Orders }
No 12   } Extract

II 2nd Lieut F. S. Calhoun 14" Infty, is hereby relieved from duty with "C" C" and will report to his proper Company Commander, Capt Thos F. Tobey for duty.

By Order of Colonel Smith
"(sgd)" d. Wm. C. McCammon
1st Lt & Adjt 14" Infty & Post.

Rec'd Feb 1 1876
F.S. Calhoun
2d Lt 14 Infty

Headquarters Camp Douglas U.T.
January 19th 1876

Special Orders
No 8
    A Garrison Court Martial is hereby ordered to Convene at this post at 10 O'Clock the 20th inst. or as soon thereafter as practicable for the trial of such Prisoners as may be properly brought before it.

 Detail for the Court
Capt Thos F Tobey
1st Lieut J.E. Quentin   14th Inftry
2nd Lieut FS. Calhoun

       By order of Col Smith
       sgd Wm W McCammon
          1st Lt & Adjt 14th Inftry & Post.

Headquarters Camp Douglas U.T.
February 1st 1876

Special Orders
No 12 -   Extract
      II 2nd Lieut F.S. Calhoun 14th Inftry is hereby relieved from duty with Co. "C" and will report to his proper Company Commander Cpt. Thos F. Tobey, for duty

       By Order of Colonel Smith
       "Sgd" Wm W McCammon
       1st Lt & Adjt 14th Inftry & Post

[In red pencil:] recd Feb 1, 1876
McCammon
14th Infantry

Headquarters Dept of the Platte
Omaha Nebraska February 17", 1876

Special Orders
No 20

A General Court Martial is hereby appointed to meet at Camp Douglas, M. T., on the 28" day of February, 1876, or as soon thereafter as practicable, for the trial of such prisoners as may be brought before it.

Detail for the Court
1 Major Montgomery Bryant 14" Infantry
2 Captain Gilbert S. Carpenter   "   "
3 1" Lieut Samuel McConihe   "   "
4 1" Lieut Julius E. Questius   "   "
5 2" Lieut Charles F. Lloyd   "   "
6 2" Lieut Frederic S. Calhoun   "   "
7 2" Lieut Joseph H Gustin   "   "

Second Lieutenant John Murphy, 14" Infantry Judge Advocate.
No other officers than those named can be assembled without manifest injury to the service —

By order of Brigadier Genl Crook
Com^{dg} the Dept of the Platte
"Sig"   R Williams
    ass^t adj^t Genl.

Headquarters Dep't of the Platte
Omaha Nebraska February 17th, 1876

Special Orders
No 210

A General Court Martial is hereby appointed to meet at Camp Douglas U.T., on the 28th day of February 1876, or as soon thereafter as practicable, for the trial of such prisoners as may be brought before it. / Detail for the Court

1 Major Montgomery Bryant 14th Infantry
2 Captain Gilbert S. Carpenter 14th Infantry
3 1st Lieut Samuel McConihe 14th Infantry
4 1st Lieut Julius E. Quentin 14th Infantry
5 2nd Lieut Charles F. Lloyd 14th Infantry
6 2nd Lieut Frederic S. Calhoun 14th Infantry
7 2nd Lieut Joseph H Gustin 14th Infantry

Second Lieutenant John Murphy, 14th Infantry Judge Advocate.

No other officers than those named can be assembled without manifest injury to the service

By order of Brigadier Genl Crook
Comdg the Dept of the Platte
"sgd" R Williams
Asst Adjt General

Headquarters Camp Douglas U.T.
March 6" 1876.

Special Orders
No 32

A Garrison Court Martial to consist of Captain James Kennington, 1" Lieut Sam'l McConihe, 2" Lieut F.S. Calhoun, 14" Infty is hereby ordered to convene at this Post at 10. O'Clock AM the 7th inst. or as soon thereafter as practicable for the trial of such prisoners as may be properly brought before it. —

By order of Col Smith

"Ofs" W.W. McCammon
1 Lieut & Adjt 14" Infty & Post.

Headquarters Camp Douglas U.T.
April 5" 1876

Special Orders
No 52

Lieut F.S. Calhoun 14" Infty, will proceed to Fort Cameron Utah on Thursday the 6" inst in charge of the Recruits of Companies 14" Infty stationed at that Post.

Having completed the duty prescribed in this Order, Lieut Calhoun will rejoin his proper station. —

The Recruits will be furnished with Eight (8) days rations. —

The Quartermaster Department will furnish the necessary transportation for One Officer and 26 Recruits. —

III Private Scharnow Co G. 14" Infantry awaiting transportation to his Company, will report to Lieut F.S. Calhoun 14" Infty

Headquarters Camp Douglas U.T.
March 6th 1876

Special Orders
No 32   A Garrison Court Martial to
Consist of Captain James Kensington
1st Lieut Saml McConihe   14th Inftry
2nd Lieut FS. Calhoun
is hereby ordered to convene at this
Post at 10 O'Clock AM the 7th inst. , or as
soon thereafter as practicable for the trial
of such prisoners as may be properly
brought before it.
            By order of Col Smith
      "sgd" Wm W McCammon
            2st Lieut & Adjt 14th Inftry & Post.

Headquarters Camp Douglas U.T.
April 5th 1876

Special Orders
No 52
         Lieut F.S. Calhoun 14th Inftry, will
proceed to Fort Cameron Utah on Thursday
the 6th inst. in charge of the Recruits
of Companies 14th Inftry stationed at that
Post.
  Having completely the duty prescribed
in this Order, Lieut Calhoun will rejoin
his proper station.
  The Recruits will be furnished with
Eight (8) days rations.
The Quartermaster Department will furnish
 the necessary transportation for One officer and 26 recruits.
  III Private Scharmon Co G. 14th Infantry awaiting
transportation to his Company, will
report to Lieut F.S. Calhoun 14th Inftry

for duty with Recruits —
Private Scharmen will be furnished with
Eight (8) days rations, and the Q. M. Dept.
will furnish necessary transportation.
By Order of Maj Bryant
Wm W McCammon
1st Lieut & Adjt
14" Infty & Post.

Left Camp Douglas in Compliance with this Order April 19"1876 & arrived at Fort Cameron April 23" 1876

Hdqrs Fort Cameron U.T.
April 26" 1876.

Special Orders
No 46 } Ex II

The Quartermasters Department will furnish transportation from this Post to Salt Lake City, U.T. for 2nd Lieut. F.S. Calhoun, 14" Infantry; he having completed the duty to which he was assigned by S. O. #52 Camp Douglas U.T. April 5" 1876.

By Order of Capt Krause
Robt A Lovell
2nd Lieut 14" Infty
Post Adjt.

Left Fort Cameron for Camp Douglas April 27" & arrived April 29" 1876.

for duty with recruits.
Private Scharmon will be furnished with
Eight (8) days rations, and the Q.M. Dept,
will furnish necessary transportation.
      By Order of Maj Bryant
  "Sgd" Wm W McCammon
      1st Lieut & Adjt
       14th Inftry & Post
[In red pencil:] Left Camp Douglas in Compliance with this order April 17th 1876 & arrived at Fort Cameron April 23rd 1876

     Hdqrs Fort Cameron U.T.
      April 26th 1876
Special Orders
No 46 -   Ex II
    The Quartermasters
Department will furnish transportation
from this Post to Salt Lake City U.T. For
2nd Lieut FS. Calhoun, 14th Infantry; he
having completed the duty to which
he was assigned by S.O. #52
Camp Douglas U.T. April 5th 1876
     By order of [illegible] Krause
  sgn" Robt A Lovell
     2nd Lt 14th Inftry Post Adjt

[In red pencil:] Left Fort Cameron for Camp Douglas April 27th & arrived April 29th 1876.

24

Headquarters Camp Douglas U.T.
May 8, 1876.

Special Orders No. 74 } A Garrison Court Martial to consist of
Captain D. W. Burke
1st Lieut. Saml. McConihe } 14" Infty
2nd Lieut. F. S. Calhoun
is hereby ordered to convene at this Post at 10 O'clock A.M. the 9" inst. or as soon thereafter as practicable for the trial of such prisoners as may be properly brought before it.

By Order of Colonel Smith
"Md" Wm. W. McCammon
1st Lieut & Adjt 14" Infty & Post.

Headquarters Camp Douglas U.T.
June 3rd 1876.

Special Orders No. 87 } A Garrison Court Martial to consist of
Captain Saml. McConihe
1st Lieut. John Murphy } 14" Infty
2nd Lieut. F. S. Calhoun
is hereby ordered to convene at this Post at 10 O'clock A.M. the 5" inst., or as soon thereafter as practicable for the trial of such prisoners as may be properly brought before it.

By Order of Col. Smith
"M" Wm. W. McCammon
1st Lieut & Adjt 14" Infty & Post.

Headquarters Camp Douglas UT.
May 8, 1876

Special Orders
No 74   A Garrison Court Martial to Consist of
Captain D.W. Burke
1st Lieut Saml McConihe   14th Inftry
2nd Lieut FS. Calhoun
is hereby ordered to convene at this
Post at 10 O'Clock A.M. the 9th inst. : or as
soon thereafter as practicable for the
trial of such prisoners as may be
properly brought before it.
         By order of Colonel Smith
         "Sgd" Wm W McCammon
         1st Lieut & Adjt 14th Inftry & Post.

Headquarters Camp Douglas U.T.
June 3rd 1876

Special Orders
No 87   A Garrison Court Martial to Consist of
Captain Saml McConihe
1st Lieut John Murphy   14th Inftry
2nd Lieut FS. Calhoun
is hereby ordered to convene at this
Post at 10 OClock AM. the 5th inst. , or
as soon thereafter as practicable for
the trial of such prisoners as may be
properly brought before it.
         By order of Col Smith
    "Sgn"   Wm W McCammon
         1st Lieut & Adjt 14th Inftry & Post

25

June 24th 1876.
Left with Company to join Big Horn & Yellowstone Expedition — No Copy of the Order received. to be copied below when recd. Fred S Calhoun

June 24th 1876
Left with Company [to join?] Big Horn & Yellowstone
Expedition - No Copy of the Order received
to be copied below where read. Fred S. Calhoun

~~In the field, Big Horn & Yellowstone Expedition~~
~~Camp Near Custer City, D.T.~~
~~Sept 23d 1876.~~

Hd Qrs Infty Battalion Big Horn & Yellowstone Expd
Camp Whitewood Creek D.T.
Sept 16° 1876.

General Orders }
No 3

The undersigned hereby relinquishes command of the Infantry Battalion, Big Horn and Yellowstone Expedition, and thanks the Officers and men of the Battalion for their promptness in the discharge of duties, their endurance of fatigue, Exposure and hunger during a march of forty days, a portion of which is unparalleled in the annals of history —

"Sgd" Alex Chambers
Major 4th Infantry
Com'dg

[First entry crossed out:]
In the field, Big Horn & Yellowstone Expedition
Camp near Custer city D.T.
Sept 23rd 1876

Hd Qrs Infty Batttalion Big Horn Yellowston Expd
Camp Whitewood Creek D.T.
Sept 16th 1876

General Orders
No 3   The undersigned hereby
relinquishes command of the
Infantry Battalion, Big Horn and
Yellowstone Expedition, and thank
the Officers and men of the
Battalion for their promptness in the
discharge of duties, their endurance
of fatigue, Exposure and hunger
during a march of forty days,
a portion of which is unparalleled
in the annals of history -

"Sgd" Alex Chambers
Major 4th Infantry
Comdg

Headquarters Big Horn & Yellowstone Exp'd
Camp near Custer City D.T.
October 15" 1876.

Special Orders }
No 9

A Board of Survey to Consist of
Captain James Kennington   14" Infty
1"Lieut W.L. Carpenter   9"   "
2"Lieut Fredk Calhoun   14"   "

will assemble at as early an hour as practicable to report upon the following Subsistence Stores, for which 1"Lieut John W. Bubb 4" Infantry A.C.S. is responsible, viz;

2000 lbs Hard Bread. 25# Flour. 24½# Cheese - 13½# Breakfast Bacon. 1 Can Fresh Pears - 1 Can Tomatoes - 1 Can Yeast Powder - 1 Bot Wor' Sauce.

By Order of Lieut Col Royall
"sgd" H R Lemly
 2nd Lieut 3" Cavalry
 A.A.A. General.

Hd. Qrs Department of the Platte
Omaha Nebraska Novem 9" 1876

Special Orders }
No 150   } Extract   *   *   *

5. Leave of absence for one month, with permission to apply for an extension of not to exceed one month, is hereby granted Second Lieutenant Fred S. Calhoun 14" Infty. (Camp Robinson)

By Command of Brig Gen Crook
 Robert Williams
"sgd"   asst adjt general.

Headquarters Big Horn & Yellowstone Expd
Camp Near Custer City D.T.
October 15th 1876

Special Orders
No 9   A Board of Survey to
Consist of
Captain James Kensington 14th Inftry
1st Lieut W.L. Carpenter 9th Inftry
2nd Lieut Fred'k Calhoun 14th Inftry
will assemble at as early an hour
as practicable to report upon the
following Subsistence Stores, for which
1st Lieut John W Bubb 4th Infantry
A.C.S. is responsible, Vizi
2000 lbs Hard Bread. 25# Flour.
24 1/2 # Cheese - 13 1/2# Breakfast Bacon,
1 Can Fresh Pears - 1 Can Tomatoes -
1 Can Yeast Powder -1 Bot Wos Sauce.[bottle Worcestershire]
                By order of Lieut Col Royall
                "sgd" H.R. Lewly
                2nd Lieut 3rd Cavalry
                A.A.A. General

[New Entry, whole entry is crossed out with the words "See Page 31" written over it in red pencil:]

                Hd Qrs Department of the Platte
                        Omaha Nebraska Novem 9th 1876

Special Orders
No 150 -   Extract
5. Leave of Absence for one month
with permission to apply for an extension
of not to exceed one month, is hereby
granted Second Lieutenant Fred S. Calhoun
14th Inftry. (Camp Robinson)
                By Command of Brig Gen Crook
    "sgd"   Robert Williams
                        Asst Adjt General

28

Head Qrs Dept. of the Platte
In the field Camp Robinson Neb
October 24" 1876

General Orders }
No 8 }

The time having arrived when the troops composing the Big Horn and Yellowstone Expedition are about to separate, The Brigadier General Commanding addresses himself to the Officers and men to say:

"In the Campaign now closed he has been obliged to call upon you for much hard service, and many sacrifices of personal comfort. At times you have been out of reach of your base of supplies in most inclement weather you have marched without food and slept without shelter — In your Engagements you have evinced a high order of discipline and courage; In your marches wonderful power of endurance, and in your deprivations and hardships, patience and fortitude —

Indian Warfare is of all warfare the most dangerous, the most trying and the most thankless: Not recognized by the high authority of the U.S. Congress as War, it still possesses for you the disadvantages of civilized warfare with all the horrible accompanyments that barbarians can invent and savages can create — In it you are required

Custer's Other Brother-in-Law

        Head Qrs Dept. of the Platte
       In the field Camp Robinson Neb
           October 24th 1876

Special Orders
No 8   The time having arrived when
the troops composing the Big Horn
and Yellowstone Expedition are about
to separate, The Brigadier General
Commanding addressees himself to the
Officers and Men to say:
"In the Campaign now closed he
has been obliged to call upon you
for much hard service, and many
sacrifices of personal comfort.
At times You have been out of reach
of your base of supplies in most
inclement weather You have
marched without food and slept
without shelter - In Your
Engagements You have evinced a
high order of discipline and
courage: In your marches wonderful
power of endurance, and in your
deprivations and hardships, patience
and fortitude -
Indian Warfare is
of all warfare the most dangerous,
the most trying and the most
thankless; Not recognized by
the high authority of the U.S. Congress
as War, it still possesses for
you the disadvantages of Civilized
Warfare with all the horrible
accompaniments that barbarians
can invent and savages can
create - In it You are recognized

to serve without the incentive to promotion or recognition; in truth without favor or hope of reward –

The people of our sparsely settled frontier in whose defense this war is waged, have but little influence with the powerful communities in the east; their Representatives have but little voice in our National Councils, while your savage foes are not only the wards of the Nation, supported in idleness, but objects of sympathy with large numbers of people, otherwise well informed and discerning –

You may therefore congratulate yourselves that in the performance of your military duty you have been on the side of the weak against the strong, and that the fair people there and on the frontier will remember your efforts with gratitude.

If in the future it should transpire that the avenues for recognition for distinguished service and gallant conduct are opened, those rendered in the Campaign will be recommended for suitable rewards; pending this, the following named Officers and men are mentioned as carrying on their persons honorable marks of distinction in the severe wounds they have received at the hands of the enemy – Viz;

Over

to serve without the incentive to
promotion or recognition; in truth without
favor or hope of reward -
      The people of our sparsely
settled frontier in whose defense this
war is waged, have but little influence
with the powerful communities in
the east: Their Representatives have
but little voice in our national
Councils, while Your savage foes
are not only the Wards of the
Nation, supported in idleness, but
objects of sympathy with large
numbers of people, otherwise well informed
and discerning -
      You may therefore
Congratulate yourselves that in the
performance of Your Military
duty You have been on the side
of the weak against the Strong,
and that the fair people there
and one the frontier will remember
Your efforts with gratitude.
      If in the future
it should transpire that the avenues
for recognition for distinguished
service and gallant conduct are
opened, those [illegible, rendered?]
in the Campaign will be recommended
for suitable rewards: Pending this,
the following named Officers and men
are mentioned as carrying on
their persons honorable marks of
distinction in the severe wounds
they have received at the hands
of the Enemy - Vis: Over

Capt	G. V. Henry	3" Cav		
Lieut	Von Ludwitz	3" Cav		
1 Sergt	Thomas Meagher	Co I	2" Cav	
Sergt	Patk O'Donnell	" D	"	
"	Andrew Grosch	I.	3" Cav	
"	Samuel Cook	L	"	"
Trumpter	Wm H. Edwards	"	"	"
Private	Henry Steiner	B	"	"
"	Chas W. Stewart	I	"	"
"	Wm Featherly	"	"	"
"	James O'Brien	"	"	"
"	Franz Smith	"	"	"
"	John Lascombeskie	"	"	"
"	John Creamer	L	"	"
"	E. A. Snow	M	"	"
"	Horace Harold	E	"	"
"	Thos Town	F	"	"
"	John H. Terry	S	4"	Infty
"	James H. Devine	D	"	"
"	Richard Flynn	"	"	"
Sergt	Ewd Schneider	K	5" Cav	
"	Ewd Glass	E.	3"	"
"	Jno H. Kirkwood	M.	3"	"
Private	J.W. Stephenson	I	2"	"
"	Wm H. Duboise	C	3"	"
"	Chas Foster	D	"	"
"	Ewd McKernan	E	"	"
"	Augustus Dorn	D	"	"
"	Ewd George Clentes	D.	5	"
"	Wm Madden	M.	5	"
"	Daniel Ford	F.	5	"
"	Michael Donally	"	"	"
"	Robt Fitz Henry	H.	9"	Infty

By Command of Brig Genl George Crook
"Sgd" John G. Bourke
1st Lieut 3" Cav A.D.C.
a.a.d. General

Capt	G.V. Henry	3"Cav	
Lieut	Von Luettwitz	3" Cav	
1st Sergt	Thomas Meagher	Co I. 2" Cav	
Sergt	Pat'k O'Donnell	" D "	
"	Andrew Loosch	I. 3" Cav	
"	Sammuel Cook	L " "	
Trumpeter	WmH. Edwards	" " "	
Private	Henry Steiuer	B " "	
"	Chas W. Stewart	I " "	
"	Wm Featherly	" " "	
"	James O'Brien	" " "	
"	Fran s Smith	" " "	
"	John Lascombeskie	" " "	
"	John Creamer	L " "	
"	E. A. Snow	M " "	
"	Horace Harold	E " "	
"	Tho s Town	F " "	
"	John H. Terry	G 4" Infty	
"	James H. Devine	D " "	
"	Richard Flynn	" " "	
Sergt	Ew'd Schneider	K 5" Cav	
"	Ew'd Glass	E. 3" "	
"	Scott Kirkwood	M. 3" "	
Private	J. W. Stephenson	I 2" "	
"	Wm H. Duboise	C 3" "	
"	Chas Foster	D " "	
"	Ew'd McKernan	E " "	
"	Augustus Dorn	D " "	
"	Ew'd George Cleutes	D. 5 "	
"	Wm Madden	M. 5 "	
"	Daniel Ford	F. 5 "	
"	Michael Donally	" " "	
"	Rob't Fitzhenry	H 9" Infty	

By command of Brig Gen'l George Crook
"sgd" John G Bourke
   1st Lieut 3rd Cav A.D.C.
   A.A.A. General

Hd Qrs Camp Robinson Neb. Oct 30" 76.

Special Orders }
No 173 }   Company "F" 14" Infantry
is hereby detailed on Special duty
at Red Cloud Indian Agency,
and will report at once to Captain
Thomas F. Tobey Actg Ind Agent:
furnishing these Hd Qrs with monthly
returns of Co. at the end of each
month
   By Order of Major J. W. Mason
"Sgd"   R. T. Yeatman
   2" F 14" Infty, Post Adjutant

Head Quarters Department of the Platte
   Omaha Nebraska Novem 9, 76
Special Orders }
No 150 } Extract. x x x x x x
5. Leave of absence for One Month,
with permission to apply for an extension
of not to exceed one Month, is hereby
granted Second Lieutenant Fred. S. Calhoun
14" Infantry. (Camp Robinson.)
   By Command of Brig- Genl Crook
"Sgd"   Robert Williams
   Asst Adjt General.

Hd Qrs Camp Robinson Neb. Oct 30th 76
Special Orders
No 173
         Company "F" 14th Inftry
is hereby detailed on special duty
at Red Cloud Indian Agency,
and will report at once to Captain
Thomas F. Tobey Actg Ind Agent:
furnishing these HdQrs with monthly
returns of Co. at the end of each
Month
    By order of Major J.W. '
    "Sgd" R.T. Yeatman
         2nd Lt 14th Inftry Post Adjutant

      Head Quarters Department of the Platte
        Omaha Nebrask Novem 9, 76
Special Orders
No 150 -
Extract   X   X   X   X   X   X
5. Leave of absence for one month
with permission to apply for an extension
of not to exceed one month, is hereby
granted Second Lieutenant Fred. S. Calhoun
14th Infantry. (Camp Robinson.)
    By Command of Brig Genl Crook
    "Sgd"   Robert Williams
        Asst Adjt General

32

Headquarters Mil. Div. of the Missouri
Chicago, Ills. December 8, 1876.

Special Orders }
No 144 } Extract.

I The leave of absence granted
2nd Lieut Fred S. Calhoun, 14th Infantry
(Camp Robinson, Neb.) by par. 5 S.O.
No 150 dated Hd. Qrs Dept of the
Platte, November 9th 1876, is hereby
extended One (1) Month.

By Command of Lieut Genl Sheridan
"signed" R. C. Drum
Asst Adjt General

Head Qrs Camp Robinson Neb
February 15" 1877

Special Orders }
No 40 (40) }
A Board of Survey is hereby
appointed to meet at this Post
at 10. A.M. tomorrow or as soon
thereafter as practicable to examine
into and determine the responsibility
of a certain lot of Ordnance,
Ordnance Stores, P. M. Property, and
Q. & S. Equipage taken by Deserters
and for which 2nd Lieut James
F. Simpson 3rd Cavalry is responsible.
Detail for the Board
Captain James Kennington   14" Infty
Captain P. D. Vroom   3" Cav
2" Lieut F. S. Calhoun   14" Infty
"Over"

Headquarters Mil Div of the Missouri
Chicago Ils Deceber 8. 1876
Special Orders
No 144 - Extract
I The leave of Absence granted
2nd Lieut Fred S. Calhoun 14th Infantry
(Camp Robinson, Neb.) by par, 5 S.O.
No 150 dated Hd.Qrs Dept of the
Platte, November 9th 1876, is hereby
extended One (1) month
By Command of Lieut Genl Sheridan
  "Signed"   R.C. Down
                Asst Adjt General

Head Qrs Camp Robinson Neb
February 15th 1877
 Special Orders
No 40
A Board of Survey is hereby
appointed to meet at this Post
at 10 AM tomorrow or as soon
thereafter as practicable to examine
into and determine the responsibility
of a certain lot of Ordnance, Ordnance
Stores, Q.M Property and
O. & G. Equipage. taken by Deserters
and for which 2nd Lieut James
F. Simpson 3rd Cavalry is responsible.
         Detail for the Board
Captain James Kensington 14th Inftry
Captain P.D. Vroom 3rd Cav
2nd Lieut F.S. Calhoun 14th Inftry

                                    "Over"

By Order of Captain D.W. Burke
"Signed" C.F. Lloyd
2nd Lieut 14" Infantry
Post Adjutant

Headquarters Camp Robinson Neb
February 17, 1877

Special Orders }
No 42 } Extract.
III 2nd Lieut J.F. Cummings 3rd Cav is hereby relieved from duty as Member of Garrison Court Martial convened by virtue of S.O. No 30 C.S. from these Hd.qrs and 2nd Lieut F.S. Calhoun 14" Infantry detailed in his stead.
By order of Captain D.W. Burke
"Sgd" C.F. Lloyd
2nd Lieut 14" Infantry
Post Adjutant.

Headquarters Camp Robinson Neb
March 5" 1877

Special Orders }
No 55 } Ex
II Lieut F. Calhoun 14" Infty, will proceed without delay to Fort Laramie W.T. for the purpose of bringing to this Post the property belonging to the Companies of the 14" Infty, which was left at Fort Laramie for storage. The Pr. M. Dept will furnish Lieut Calhoun with one (1) Pack Mule Team and one (1) Saddle Horse.
By order of Capt D.W. Burke
"Sgd" C.F. Lloyd
2nd Lieut 14" Infantry
Post Adjutant

By order of Captain D.W. Burke
"Signed" CF. Lloyd
2nd Lieut 14th Infantry
Post Adjutant

Headquarters Camp Robinson Neb
February 17, 1877

Special Orders
No 42 - Extrac
II   2nd Lieut J.F. Cummings 3rd Cav
is hereby relieved from duty as
member of Garrison Court Martial
convened by virtue of S.O. No 30
C.F. from these HdQrs and 2nd Lieut.
FS. Calhoun 14th Infantry detailed
in his stead.

By order of Captain D.W. Burke
"Sgd"   C.F. Lloyd
2nd Lieut 14th Infantry
Post Adjutant

Headquarters Camp Robinson Neb
March 5th 1877

Special Orders
No 55 - Ex
II Lieut F. Calhoun 14th Inftry, will proceed
without delay to Fort Laramie W.T. for the
purpose of bringing to this Post the property
belonging to the Companies of the 14th Inftry, which
was left at Fort Laramie for storage. The
Qr Ms. Dept will furnish Lieut. Calhoun with
One (1) six mule team and one (1) saddle
Horse.

By order of Capt D.W. Burke
"sgd" C.F. Lloyd
2nd Lieut 14th Infantry
Post Adjutant

34

Headquarters Camp Robinson Neb
March 21st 1877

Special Orders } Ex.
No 69

VII Second Lieutenant F.S. Calhoun 14" Infantry will report at 9 oclock A.M. tomorrow, to the Acting Indian Agent, Red Cloud Agency, Neb, to receive and inspect a quantity of Beef for issue to Indians.

By Order of Colonel R.S. Mackenzie
"sgd" Joseph H. Dorst
2nd Lt & act'g Adj't 4" Cav'y
Post Adjutant.

Headquarters Camp Robinson Neb
March 23rd 1877

Special Orders }
No 71

III A Board of Survey is hereby ordered to convene at this Post at 10 oclock A.M. tomorrow or as soon thereafter as practicable to examine into and ascertain what articles of Government property have been lost or abstracted by Private Crist Abell, a deserter from Co "D" 14" Infantry, and for which Captain D.W. Burke 14" Inf'ty is responsible.

Detail for the Board.
Captain James Kennington   14" Inf'ty
2" Lieut C.F. Lloyd   14" Inf'ty
2" Lieut F.S. Calhoun   14" Inf'ty

By Order of Col R.S. Mackenzie
"sgd" Joseph H. Dorst
2nd Lt & Act. Adj't 4" Cav'y
Post Adjutant.

Headquarters Camp Robinson Neb

March 21st 1877

Special Orders
No 69 -   Ex
VII Second Lieutenant F.S. Calhoun
14th Infantry will report at 9 O'Clock A.M.
tomorrow, to the Acting Indian
Agent, Red Cloud Agency, Neb, to
receive and inspect a quantity of Beef
for issue to Indians.
  By Order of Colonel R.S. Mackenzie
  "Sgd" Joseph H Durst
     2nd Lt & Acting Adjt 4th Cav'y
       Post Adjutant.

Headquarters Camp Robinson Neb
    March 23rd 1877

Special Orders
No 71
 IIII A Board of Survey is hereby
ordered to convene at this Post
at 10 OClock A.M. tomorrow or
as soon thereafter as practicable
to examine into and ascertain
what articles of Government
property have been lost or abstracted
by Private Crist[?] Abell, a Deserter from
Co "C" 14th Infantry, and for which
Captain D.W. Burke 14th Inftry
is responsible.
  Detail for the Board
Captain James Kensington 14th Inftry
2nd Lieut C.F. Lloyd 14th Inftry
2nd Lieut F.S. Calhoun 14th Inftry
   By order of col R.S. Mackenzie
   "sgd" Joseph H Dorst
   2nd Lt & Act Adjt 4th Cavalry
    Post Adjutant

Headquarters Camp Robinson Neb
March 26" 1874.

Special Orders }
No 73 }

II A Garrison Court Martial will convene at this Post at 10. O'ck A.M. tomorrow or as soon thereafter as practicable, for the trial of such prisoners as may be properly brought before it.

Detail for the Court.
Captain Thos F. Tobey   14" Infty
1" Lieut W.L. Carpenter   9" Infty
2" Lieut F.G. Calhoun   14" Infty

By order of
Col R.S. Mackenzie
"Sgd" Joseph H. Dorst
2" Lieut & act adjt 4" Cav'y
Post Adjutant.

Headquarters Camp Robinson Neb
April 9" 1874.

Special Orders }
No 81 } Extract.

III A Board of Survey is hereby ordered to assemble at this Post at 10 O'clock A.M. tomorrow the 10" inst, to ascertain what articles of Government property, have been lost or abstracted by Private Theodore F. Johnson Co I, 9" Infantry, who deserted on the 7" inst while on detached service at Hat Creek Wyo.

Detail for the Board
Capt James Kennington   14" Infantry
2" Lt C.F. Loyd   14" Infty
2" Lt F.G. Calhoun   14" Infty
                          "Over"

Headquarters Camp Robinson Neb
March 26th 1877

Special Orders
No 73
II A Garrison Court Martial will convene
at this post at 10 Ock A.M. Tomorrow
or as soon thereafter as practicable,
for the trial of such prisoners as
may be properly brought before it.
     Detail for the Court.
 Captain Thos F. Tobey 14th Inftry
 1sst Lieut W.L. Carpenter 9th Inftry
 2nd Lieut F.S. Calhoun 14th Inftry
  By order of Col R.S. MacKenzie
  "Sgd" Joseph H. Dorst
   2nd Lieut & act adjt 4th Cavly
    Post Adjutant

   Headquarters Camp Robinson Neb
    April 9th 1877

Special Orders
No 84 - Extract.
III A Board of Survey is hereby
ordered to assemble at this Post
at 10 O'clock A.M. tomorrow the
10th inst., to ascertain what articles
of Government property, have been
lost or abstracted by Private Theodore
F. Johnson Co G. 9th Infantry, who
deserted on the 7th inst. while on
detached service at Hat Creek
Wy.
    Detail for the Board
Capt James Kensington 14th Infantry
2nd Lt C.F. Lloyd 14th Inftry
2nd Lt F.S. Calhoun 14th Inftry   "Over"

The proceedings will be rendered agreeably to the closing paragraph of General Orders No 110 Hd Qrs of the Army, series of 1876.

By order of
Col. R. S. Mackenzie
"Sgd" Joseph H. Dorst
2nd Lt & Act Adjt 4" Cav
Post Adjutant.

Headquarters Camp Robinson Neb
April 14" 1877.

Special Orders } Ex
No 92 }

VIII A Board of Survey will be convened at this post at 1. OCk P.M. this day or as soon thereafter as practicable, to enquire into and fix the responsibility for the loss of one (1) box of Pears (24 Cans) invoiced by Major John P. Hawkins C.S. on the 13" of October 1876 to 1" Lieut John Murphy 4" Infantry A.C.S. but not yet received.

Detail for the Board.
Captain James Kennington   14" Infty
2" Lt   C. F. Lloyd   "   "
2" Lt   F. S. Calhoun   "   "

By order of
Captain C. Mauck
"Sgd" C. F. Lloyd
2" Lt 14" Infty
Acty Post Adjt.

The proceedings will be rendered
agreeably to the Closing paragraph
of General Orders No 110 HdQrs
of the Army, series of 1876.
    By order of
        Col R.S. Mackenzie
"Sgd" Joseph H. Dorst
      2nd Lt & Act Adjt 4th Cav
           Post Adjutant.

                Headquarters Camp Robinson Neb
                      April 14th 1877
Special Orders
No 92 -   Ex
III A Board of Survey will be
convened at this post at 1 Ock
P.M. this day or as soon thereafter
as practicable to enquire into
and fix the responsibility for
the loss of One (1) box of Pears (24
cans) invoiced by Major John
P. Hawkins C.S. on the 13th of October
1876 to 1st Lieut John Murphy 14th
Infantry A.C.S. but not yet received.
    Detail for the Board.
Captain James Kensington 14th Inftry
2nd Lt CF. Lloyd 14th Inftry
2nd Lt F.S. Calhoun 14th Inftry
      By order of
           Captain C. Mauck[?]
    "Sgd" C.F. Lloyd
        2nd Lt 14th Inftry
          Actg Post Adjt.

37

Headquarters Camp Robinson Neb
May 3rd 1877

Special Orders)
No 109)

A Board of Survey will convene at this post at 10. o'clock A.M. tomorrow, or as soon thereafter as practicable, to inquire into, and report upon the circumstances attending, and to fix the responsibility for the loss of One Wall Tent and One set Wall Tent Poles, the property of the United States and for which Captain Thomas F. Tobey 14th Infantry, is responsible.

Detail for the Board:
Captain James Kennington 14" Infty
1st Lieut W. L. Carpenter 9" Infty
2d Lieut F. J. Calhoun 14" Infty

By order of Col R.S. Mackenzie
"Sgd" Joseph H. Dorst
2d Lt & Actg Adjt 4" Cavalry
Post Adjutant.

Hd Qrs Camp Robinson Neb
August 9" 1877

Special Orders)
No 190) 1st Lieut C. A. Johnson 14" Infantry is hereby detailed to be present at all issues to Indians, at Red Cloud Indian Agency, and also to witness the receipt of Cattle and all stores intended as supplies for Indians Vice 2d Lieut F.J. Calhoun 14" Infty who is hereby relieved

By order of Lt Col L.P. Bradley
"Sgd" C. H. Floyd 2nd Lieut 14" Infty Post Adjt.

118

Headquarters Camp Robinson Neb
May 3rd 1877

Special Orders
No 109

A Board of Survey will convene at this post at 10 O'clock A.M. Tomorrow or as soon thereafter as practicable, to inquire into, and report upon the circumstances attending, and to fix the responsibility for the loss of One Wall Tent and One set Wall Tent Poles, the property of the United States and for which Captain Thomas F. Tobey 14th Infantry, is responsible.

Detail for the Board:
Captain James Kensington 14th Inftry
1st Lieut W.L. Carpenter 9th Inftry
2nd Lieut F.S. Calhoun 14th Inftry

By order of Col R.S. Mackenzie
"sgd" Joseph H Dorst
2nd Lt & Actg Adjt 4th Cavalry
Post Adjutant

[New Entry, entirely crossed out:]
HdQrs Camp Robinson Neb
August 9th 1877

Special Orders
No 190   II 1st Lieut CA Johnson 14th Infantry is hereby detailed to represent at all issues to Indians at Red Cloud Indian Agency, and also to [illegible] the receipt of Cattle and all stores intended as supplies for Indians' Vice 2d Lieut F.S. Calhoun who is hereby relieved
By order of Lt Col L.P. Bradley
Sgd CF Lloyd 2nd Lt 4th Infty Post Adjt.

Headquarters Camp Robinson Neb
May 13th 1877.

Special Orders(
No 17   Extract.

III A Board of Survey will convene at this Post at 10 o'clock A.M. tomorrow, or as soon thereafter as practicable to investigate the circumstances attending, and fix the responsibility for, the loss of three Sharps Carbines, the Property of the United States, for which Captain G. M. Randall, 23° Infantry is responsible.

Detail for the Board.
Captain James Kennington   14° Infantry.
1st Lieut W. L. Carpenter   9" Infantry.
2nd Lieut F. J. Calhoun   14° Infantry.

The Board will also investigate the circumstances of, and fix the responsibility for, the loss of one Case of Potatoes (12 Cans) the property of the United States, for which 1st Lieut John Murphy, 14th Infantry, A.C.S. is responsible, reported lost en-route between Sidney and Camp Robinson, Neb. The proceedings in each case will be rendered in triplicate.

By order of Col R. Mackenzie
    (sgd) Joseph H. Dorst
    1st Lt & Act Adjt 4° Cavalry
    Post Adjutant.

Headquarters Camp Robinson Neb
May 13th 1877

Special Orders
No 17   - Extract.
III A Board of Survey will convene
at this Post at 10 Oclock A.M.
tomorrow or as soon thereafter as
practicable to investigate the
circumstances attending, and fix
the responsibility for, the loss of
three Sharps Carbines, the Property
of the United States, for which
Captain G.M. Randall, 23rd Infantry
is responsible.

                    Detail of the Board.
Captain James Kensington 14th Infantry
1st Lieut W.L. Carpenter 9th Infantry
2nd Lieut F.S. Calhoun 14th Infantry
The Board will also investigate
the circumstances of, and fix the
responsibility for, the loss of One
Case of Tomatoes (12 cans) the property
of the United States, for which 1st Lieut
John Murphy, 14th Infantry, A.C.S. Is
responsible, reported lost en-route
between Sidney and Camp Robinson,
Neb.

  The proceedings in each case
will be rendered in triplicate
      By order of Col R. Mackenzie
        "Sgd"   Joseph H. Dorst 2nd Lt & Act Adjt 4th Cavalry
                Post Adjutant

Headquarters Camp Robinson Neb
May 29" 1877.

Special Orders (Extract.
No. 125 (

× × × × ×

VII A Board of Survey will convene at this Post at 1 O'ck P.M. this day, or as soon thereafter as practicable, to inquire into, report upon, and fix the responsibility for, the deficiency of one Sharps Carbine, found to exist between the Invoice of Major A.W. Evans, 3rd Cav'y, to 1st Lt W.P. Clark 2nd Cav'y, and the number received.

Detail for the Board
Captain James Kennington 14th Infantry
1st Lieut W.L. Carpenter 9th Infantry
2nd Lieut H.J. Goldman 14th Inf'try

By order of Col R.S. Mackenzie
Sig'd Joseph H. Dorst
2nd Lt & Act Adj't 4th Cav'y
Post Adjutant.

Headquarters Camp Robinson Neb
May 22nd 1877

Special Orders
No 125 -   Extract.
X   X   X   X
   X

III A Board of Survey will convene at
this Post at 1 O'ck P.M. this day,
 or as soon thereafter as practicable,
to inquire into, report upon, and
fix the responsibility for, the
deficiency of one Sharps Carbine,
found to exist between the Invoice of
Major A.W. Evans, 3rd Cavly, to 1st
Lt W.P. Clark 2nd Cavalry, and the
number received.

   Detail for the Board
Captain James Kensington 14th Infantry
1st Lieut W.L. Carpenter 9th Infantry
2nd Lieut F.S. Calhoun 14th Inftry
         By order of Col R.S. Mackenzie
             "Sgd" Joseph H. Dorst 2nd Lt & Act Adjt 4th Cavly
                         Post Adjutant.

40

Headquarters Camp Robinson Neb
May 29" 1877

Special Orders } Ex
No 133 }
VIII 2nd Lt F. S. Calhoun, 14" Infantry
will report in person to Captain
Randall 23" Infty, for duty.
The Post Quartermaster will furnish
him with One Government Horse,
Saddle, Saddle Blanket and bridle.
By order of Lt Col L. P. Bradley.
"Sg̃d" C. F. Lloyd
             2nd Lt 14" Infantry
             Post Adjutant.

Headquarters Camp Robinson Neb
June 16" 1877.

Special Orders }
No 150   }
II Capt Thos F. Tobey 14th Infantry is
hereby relieved and 2nd Lieut
F.S. Calhoun 14" Infantry, detailed
to be present at all issues to
Indians at Red Cloud Indian
Agency, and also to Witness
the receipt of Cattle and all
Stores intended as supplies for
Indians
By Order of Lt Col L. P. Bradley
"Sg̃d" C. F. Lloyd
             2nd Lt 14" Infty
             Post Adjutant.

> Headquarters Camp Robinson Neb
> May 29th 1877

Special Orders
No 133 -   Ex
III 2nd Lt F.S. Calhoun, 14th Infantry
will report in person to Captain
Randall 23rd Infty, for duty.
The Post Quartermaster will furnish
him with One Government Horse
Saddle, Saddle Blanket and bridle.
> By order of Lt Col L.P. Bradley
> "Sgd" C.F. Lloyd
> 2nd Lt 14th Infantry
> Post Adjutant.

> Headquarters Camp Robinson Neb
> June[?] 16th 1877

Special Orders
No 150
II Capt Thos F. Tobey 14th Infantry is
hereby relieved and 2nd Lieut
F.S. Calhoun 14th Infantry, detailed
to be present at all issues to
Indians at Red Cloud Indian
Agency, and also to Witness
the receipt of Cattle and all
Stores intended as supplies for
Indians
  By order of Lt. Col L.P. Bradley
> "Sgd" C.F. Lloyd 2nd Lt 14th Inftry
> Post Adjutant

Headquarters Camp Robinson Neb
June 29" 1877

Special Orders }
No 161

II 2no Lt F. G. Calhoun 14" Infantry,
is hereby detailed on Special duty in
Command of Co "B" 14" Infantry, during
the illness of its Captain.
By order of Lt Col L. P. Bradley
"Sgd" C. F. Lloyd
2no Lt 14" Infty
Post Adjutant.

Headquarters Camp Robinson Neb
July 5" 1877

Special Orders }
No 167

II 2nd Lieut F. G. Calhoun 14th Infantry
is hereby relieved from Special duty in
Command of Co "B" 14" Infantry to date
from July 4" 1877, and detailed on
Special duty in Command of Co "G"
9" Infantry during the temporary absence
of its Commanding Officer.
By order of Captain H. W. Wessells Jr
"Sgd" C. F. Lloyd
2nd Lt 14" Infty
Post Adjutant

Headquarters Camp Robinson Neb
June 29th 1877

Special Orders
No 161

II 2nd Lt F.S. Calhoun 14th Infantry, is hereby detailed on Special duty in Command of Co "B" 14th Infantry, during the illness of its Captain

By order of Lt Col L.P. Bradley
"Sgd" C.F. Lloyd
2nd Lt 14th Inftry
Post Adjutant.

Headquarters Camp Robinson Neb
July 5th 1877

Special Orders
No 167

II 2nd Lieut F.S. Calhoun 14th Infantry is hereby relieved from Special duty in Command of Co "B" 14th Infantry to date from July 4th 1877, and detailed on Special duty in Command of Co "G" 9th Infantry during the temporary absence of its Commanding Officer

By order of Captain H.W. Wessells Jr
"Sgd" C.F. Lloyd 2nd Lt 14th Infty
Post Adjutant

Headquarters Camp Robinson Neb
July 14th 1877.

Special Orders
No 174

II 2nd Lieut F. S. Calhoun 14th Infantry is hereby detailed on Special Duty with Company "B" 14th Infty, and will report to Captain James Kennington 14th Infty for duty.

By order of
Lieut Col L. P. Bradley
Sgd. C. F. Lloyd
2nd Lt 14th Infty
Post Adjutant.

Headquarters Camp Robinson Neb
July 19th 1877

Special Order
No 177

II 2nd Lt F. S. Calhoun 14th Infantry with Sergeant Kelly and Private Little Co F. 14th Infty will proceed without delay to Omaha Barracks Neb in charge of Prisoners Bright late Private Co C 14th Infty and J. W. Vincent late Private Co M 3rd Cavalry sentenced to Military Prison by order of G. C. M. orders No 67 & 71 Hdqrs Dept of the Platte. Upon arrival at Omaha Bks Lieut Calhoun will report and turn over the Prisoners to the Commanding Officer of that Post. The quartermasters Dept will furnish the necessary transportation.

x x x x x x x x

By order of Lt Col L. P. Bradley
Sgd. C. F. Lloyd
2nd Lt 14th Inft Post Adjt.

Headquarters Camp Robinson Neb
July 14th 1877

Special Orders
No 177
II 2nd Lieut F.S. Calhoun 14th Infantry
is hereby detailed on Special
Duty with Company "B" 14th Infty
and will report to Captain James
Kensington 14th Infty for duty.
By order of
    Lieut Col L.P. Bradley
    "Sgd" C.F. Lloyd
      2nd Lt 14th Infty
        Post Adjutant

Headquarters Camp Robinson Neb
July 19th 1877

Special Orders
No 177
II 2nd Lt F.S. Calhoun 14th Infantry with
Sergeant Kelly and Private Zitter[?] Co F. 14th
Infty will proceed without delay to
Omaha Barracks Neb in charge of Prisoners
Bright late Private Co "C" 14th Inftry and
G.W. Vincent late Private Co M. 3rd Cavalry
Sentenced to Military Prison by
order of G.C.M. orders No 67 & 71
Hdqrs Dept of the Platte. Upon arrival
at Omaha Bks Lieut Calhoun will
report and turn over the Prisoners
to the Commanding Officer of that
Post. The quartermasters Dept will
furnish the necessary transportation.
By order of Lt Col L.R. Bradley
    "sgd" C.F. Lloyd
      2nd [illegible] Post Adjt

Headquarters Camp Robinson, Neb
August 9" 1877

Special Orders
No 190

II 1" Lieut C. A. Johnson 14" Infty is hereby detailed to be present at all issues to Indians, at Red Cloud Agency, and also to witness the receipt of Cattle, and all stores intended as Supplies for Indians, vice 2d Lt F. S. Calhoun 14" Infty, who is hereby relieved

By order of
Lt Col L. P. Bradley
"Sgd" C. F. Lloyd
2nd Lt 14" Infantry
Post Adjutant

Headquarters Dept of the Platte
Omaha Nebraska July 21, 1877

Special Orders
No 95

A. General Court Martial is hereby appointed to convene at Camp Robinson Neb, on the 25" day of July 1877, or as soon thereafter as practicable, for the trial of such prisoners as may be brought before it.

Detail for the Court:
1 Capt Thos M. Burrowes 9" Infty
2 " Jas Kennington 14" "
3 " Thos F. Tobey 14" "
4 1" Lt John Murphy 14" "
5 2" Lt Fred'k S. Calhoun 14" "
6 " " Chas L. Hammond 3" Cavy

Headquarters Camp Robinson Neb
August 9th 1877

Special Orders
No 190
  II 1st Lieut C.A. Johnson 14th Inftry is hereby detailed to be present at all issues to Indians, at Red Cloud Agency, and also to witness the receipt of Cattle, and all stores intended as supplies for Indians Vice 2d Lt F.S. Calhoun 14th Inftry who is hereby relieved
    By order of Lt Col L.P Bradley
      "Sgd" CF. Lloyd
      2nd Lt 14th Infantry
      Post Adjutant

Headquarters Dept of the Platte
Omaha Nebraska July 21, 1877

Special Orders
No 95
  A. General Court Martial is hereby appointed to Convene at Camp Robinson Neb, on the 25th day of July 1877, or as soon thereafter as practicable, for the trial of such prisoners as may be brought before it.
    Detail for the Court:
1 Capt Thos M. Burrowes 9th Inftry
2 Capt Jos Kensington   14th "
3 Capt Thos F Tobey   14th "
4 1st Lt John Murphy   14th "
5 2nd Lt Fredk S. Calhoun 14th "
6 2nd Lt Chas L. Hammond 3rd Cavly

7 2nd Lt Joseph H. Cummings 3 Cavly
2nd Lt Charles F. Lloyd, 14" Infty, Judge
Advocate.
No Other Officers than those named
can be assembled without manifest
injury to the service.
By order of Brig Gen'l Crook
Cmdg Dept of the Platte
"Sgd" Robert Williams
Asst Adjt General.

Head qrs Dist of the Black Hills
Camp Robinson Neb
August 12" 1877.

General Orders
No 4

2nd Lieut F.S. Calhoun 14 th
Infantry is hereby appointed Actg
Asst Adjutant General District
of the Black Hills vice 2nd Lieut
C. F. Lloyd 14 th Infantry, who is
hereby relieved to enable him
to take advantage of Leave of Absence.
Sgd L. P. Bradley
Lieut Col 9" Infty
Commanding.

7 2nd Lt Joseph H Cummings 3 Cavly
2nd Lt Charles F Lloyd, 14th Inftry, Judge Advocate
 No other officers than those named
can be assembled without manifest
injury to the service.
  By order of Brig Genl Crook
   [illegible] Dept of the Platte
  "sgd" Robert Williams
    Asst Adjt General

   Headqrs Dep't of the Black Hills
    Camp Robinson Neb
     August 12th 1877
General Orders
No 4
  2nd Lieut F.S. Calhoun 14th Infantry
is hereby appointed Actg
Asst Adjutant General District
of the Black Hills Vice 2nd Lieut
C.F. Lloyd 14th Infantry who is
hereby relieved to enable him
to take advantage of Leave of Absence.
  sgd L.P. Bradley
   Lieut Col 9th Infty
    Commanding

Custer's Other Brother-in-Law

Head Qrs Camp Robinson Neb
August 13th 1877.

General Orders
No 59.

2nd Lt. F. J. Calhoun 14th Infty
is hereby appointed Post Adjutant,
Post Treasurer, and Act'g Signal
Officer at this Post vice 2nd Lieut
C. F. Floyd 14th Infty who is hereby
relieved to enable him to take
advantage of Leave of Absence.

"Sgd" L. P. Bradley
Lt Col 9th Infty
Cmdg Post

Headquarters Dept of the Platte
Omaha Neb August 8" 1877.

Special Orders
No 100

* * * * * *

IIIII Second Lieutenant Frederic J.
Calhoun 14th Infantry is hereby
relieved from duty as Member of the
General Court Martial convened at Camp
Robinson, Nebraska, by Special Orders
No 95 C. S. from these Hd Qrs, and
detailed as Judge Advocate of the
same Court, vice Second Lieutenant
Charles F. Floyd 14th Infty, who is hereby
relieved.

* * * * * * * *

By Command of Brig Gen'l Crook

"Sgd" Robert Williams
Asst Adj't General.

Headqrs Camp Robinson Neb
August 12th 1877

General Orders
No 59
    2nd Lt F.S. Calhoun 14th Inftry is hereby appointed Post Adjutant, Post Treasurer, and Actg Signal Officer at this Post Vice 2nd Lieut C.F. Lloyd 14th Infty who is hereby relieved to enable him to take advantage of Leave of Absence.
    "Sgd" L.P. Bradley
      Lt Col 9th Infty
        Comdig' Post

Headquarters Dept of the Platte
Omaha Neb August 8th 1877

Special Orders
No 100 - Ex   X   X   X   X   X   X
IIII Second Lieutenant Frederic S. Calhoun 14th Infantry is hereby relieved from duty as Member of the General Court Martial Command at Camp Robinson, Nebraska, by Special Orders No 95 C.S. from these HdQrs, and detailed as Judge Advocate of the same court, Vice Second Lieutenant Charles F. Lloyd 14th Inftry, who is hereby relieved.
X   X   X   X   X   X
    By Command of Brig Genl Crook
    "Sgd" Robert Williams
      Asst Adjt General

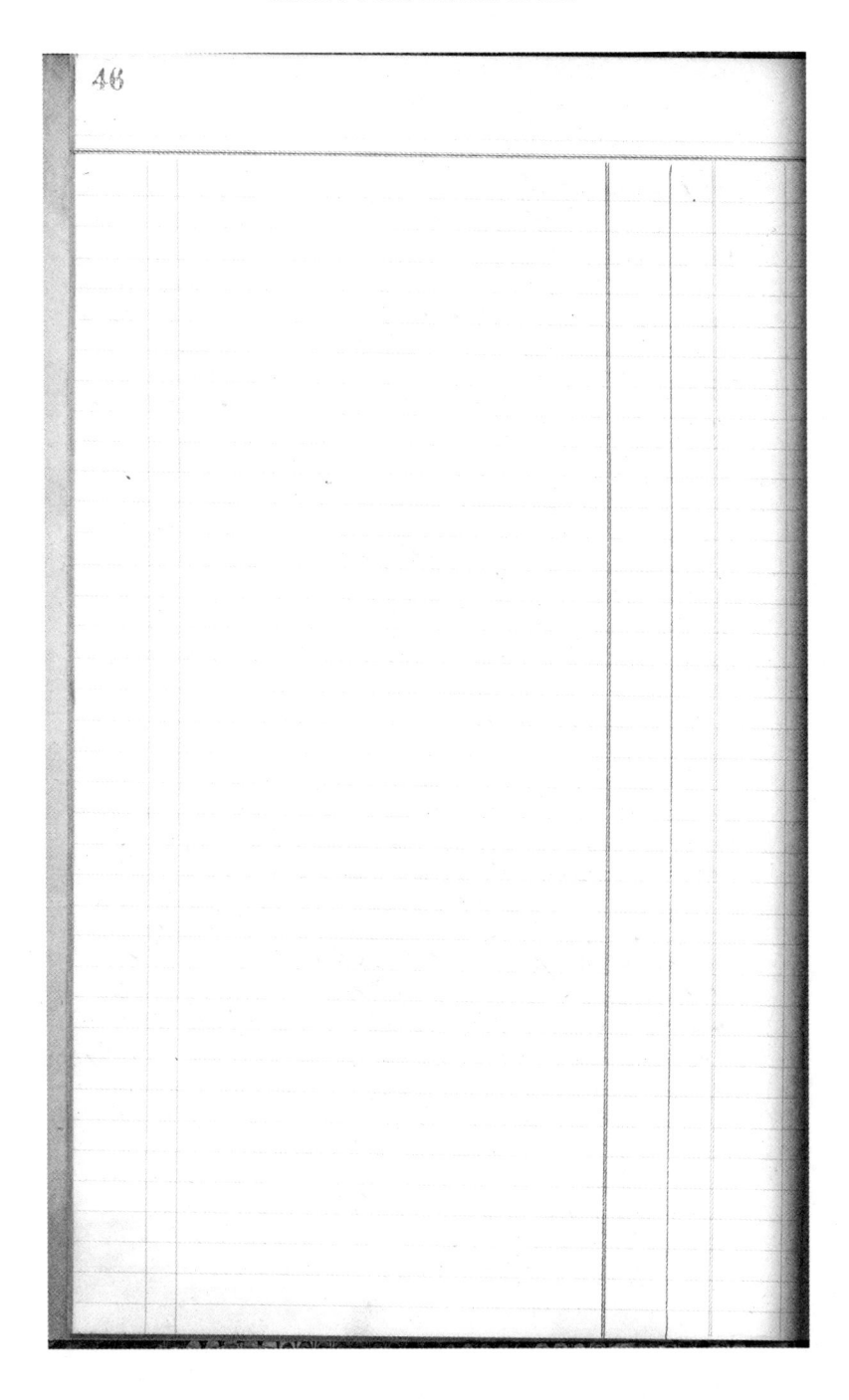

46

Custer's Other Brother-in-Law

Hd. Qrs. Dept of the Platte.
Omaha Neb Oct 26, 1877

Special Orders }
No 125   } E,

1. A Board of Survey to consist of Captain Thomas B. Burrowes, 9" Infty, Captain Thomas F. Tobey, 14" Infty, and Second Lieutenant Fred S. Calhoun, 14th Infantry, will convene at Camp Robinson Neb, on Monday the 29" day of October 1877, at 10 Oclock AM. or as soon thereafter as practicable, to examine into, report upon, and fix the responsibility for a deficiency in Ordnance Stores shipped from Rock Island Arsenal to Lieutenant Colonel L. P. Bradley, 9" Infty, Commanding Camp Robinson Neb.

+ + + + +

By Command of Brig Gen'l Crook
"Sgd"   Robert Williams
        Asst Adjt General.

Headquarters Camp Robinson Neb Nov 4, 77

Special Orders }
No 256   } E

In pursuance of S.O. No 25 dated Hd Qrs Dept of the Platte Omaha Neb Oct 26, 77 Co's G 9" Inf and B & F" 14" Inf are hereby relieved from duty at this post and will proceed to Sidney Barracks, en-route to their proper posts as soon as the Quartermaster can furnish transportation.

By order of Lt Col Bradley
"Sgd"   Bainbridge Reynolds
        2nd Lt 3" Cav Post Adj't.

Hd.Qrs Dept of the Platte
Omaha Neb Oct 26, 1877

Special Orders
No 25 -   Ex
II A Board of Survey to consist of Captain
Thomas B. Burrowes, 9th Inftry, Captain Thomas
F. Tobey, 14th Inftry and Second Lieutenant
Fred S. Calhoun, 14th Infantry, will
convene at Camp Robinson Neb on Monday
the 29th day of October 1877, at 10 Oclock
A.M. or as soon there after as practicable,
to examine into, report upon, and fix
the responsibility for a deficiency in
Ordnance Stores shipped from Rock
Island Arsenal to Lieutenant Colonel
L.P. Bradley, 9th Inftry, Commanding
Camp Robinson Neb.

X   X   X   X   X

By Command of Brig Genl Crook
  "sgd"   Robert Williams
            Asst Adjt General

Headquarters Camp Robinson Neb Nov 4, 77
Special Orders
No 256 -   Ex
 In pursuance of G.O. No 25 dated HdQrs Dept
of the Platte Omaha Neb Oct 26, 77 Cos G 9th Inf
and "B & F" 14th Inf are hereby relieved from duty at this
post and will proceed to Sidney Barracks, en-route to their
proper
posts as soon as the quartermaster can furnish
transportation.
            By order of Lt Col Bradley
            "sgd" Bainbridge Reynolds
                  2nd Lt 3rd Cav Post Adjt

48

Hd. Qrs Dept of the Platte
Omaha, Neb. Nov 12" 1877

Special Orders }
No 130 } Ex+

3. A General Court Martial is hereby appointed to meet at Camp Douglas Utah, on the 19" day of November 1877, or as soon thereafter as practicable, for the trial of such prisoners as may be brought before it.

Detail for the Court:
1 Maj Montgomery Bryant 14" Infty
2 " Henry T. Thomas 4" Infty
3 Capt Wm S. Collier 4" Infty
4 " Thos F. Tobey 14" "
5 1st Lt Frank Taylor " "
6 2nd Lt Richard T. Yeatman " "
7 2" " Robert A. Lovell " "
8 " " Fredk S. Calhoun " "
9 " " Wm A. Kimball " "
First Lieut Patrick Hasson 14" Infty
Judge Advocate.
No other Officers than those named can be assembled without manifest injury to the service.
The Court is authorized to sit without regard to hours.

By order Brig Genl Crook
Cmdg Dept of the Platte
"Sig" Robert Williams
Asst Adj't General.

Hd.Qrs Dept of the Platte
Omaha, Neb Nov 12th 1877

Special Orders
No 130 -   Ex

  X X X X

3. A General Court Martial is hereby appointed to meet at Camp Douglas Utah, on the 19th day of November 1877, or as soon thereafter as practicable, for the trial of such prisoners as may be brought before it.

Detail for the Court:

1 Maj Montgomery Bryant 14" Inftry
2 Maj Henry G. Thomas 4" Inftry
3 Capt Wm S. Collier   4" Inftry
4 Capt Thos F. Tobey   14" "
5 1st Lt Frank Taylor   " "
6 2nd Lt Richard T. Yeatman " "
7 2nd Robert A. Lovell " "
8 2nd Lt Fredk S. Calhoun " "
9 2nd Lt Wm A Kimball " "

First Lieut Patrick Hasson 14th Inftry Judge Advocate.

No other officers than those named can be assembled without manifest injury to the service.

The Court is authorized to sit without regard to hours.

X x X

By order Brig Genl Crook
 Comdg Dept of the Platte
 "Sgd" Robert Williams
  Asst Adjt General

Hd Qrs Camp Douglas U.T.
November 21" 1877

Special Orders }
No 127   } Ex
VII   A Garrison Court Martial to
Consist of
Capt Thomas F. Tobey   }
1st Lieut Patrick Hasson } 14" Infty
2d Lieut Fred J. Calhoun }
is hereby appointed to convene at
this post at 10 O'Clock A.M. the 22nd
inst, or as soon thereafter as practicable,
for the trial of such prisoners as
may be properly brought before it.
     By order of Colonel Smith
"sgd"   W W McCammon
         1st Lt & Adjt 14" Infty
              Post Adjutant.

S.O. No 136 – Substituted Lieut Murphy
for Lieut Hasson.

Hd Qrs Camp Douglas U.T.
November 21st 1877

Special Orders
No 127 -   Ex
II A Garrison Court Martial to
Consist of
Capt Thomas F. Tobey
1st Lieut Patrick Hasson   14th Infantry
2nd Lieut Fred. S. Calhoun
is hereby appointed to Convene at
this post at 10 O'Clock A.M. the 22nd
inst., or as soon thereafter as practicable,
for the trial of such prisoners as
may be properly brought before it.
By order of Colonel Smith
"sgd" Wm W McCammon
1st Lt & Adjt 14th Inftry
Post Adjutant.

S.O. No 136 - Substituted Lieut Murphy
for Lieut Hasson.

50

Hd. Qrs Camp Douglas U.T.
Dec 18. 1877

Special Orders }
No 139   } ¶1

II A Garrison Court Martial to consist of:
Maj B. A. Clements Surg. U.S.A.
1st Lt John Murphy   14" Infty
2nd Lt Fred J Calhoun 14" Infty
is hereby appointed to convene at this
post at 10 o'ck A.M. the 19" inst., or
as soon thereafter as practicable,
for the trial of such prisoners as
may be properly brought before it.

By order of Captain Tobey
"Sgd" Wm W. McCammon
1st Lt & Adj't 14" Infty
Post Adjutant.

Headquarters Camp Douglas U.T.
January 16" 1878

Special Orders }
No 6   } ¶1

III Second Lieutenant F. J. Calhoun
14" Infantry, will proceed to Corinne
Utah, to identify a civilian there
in Confinement, under charge of
Complicity in the recent theft of
Arms at this post — Upon completion
of which duty he will rejoin his
proper station —

By order of Captain Tobey
"Sgd" Wm W. McCammon
1st Lt & Adj't 14" Infty
Post Adj't.

Hd.Qrs Camp Douglas U.T.
Dec 18, 1877

Special Orders
No 139 -   Ex
II A Garrison Court Martial to Consist of:
Maj B.A. Clements Surg U.S.A.
1st Lt John Murphy 14th Inftry
2nd Lt Fred. S Calhoun 14th Inftry
is hereby appointed to Convene at this
post at 10 O'ck A.M. the 19th inst. or
as soon thereafter as practicable,
for the trial of Such prisoners as
may be properly brought before it.
    By order of Captain Tobey
     "sgd" Wm W. McCammon
      1st Lt & Adjt 14th Inftry
       Post Adjutant

Headquarters Camp Douglas U.T.
January 16th 1878

Special Orders
No 6 -   Ex
II Second Lieutenant F.S. Calhoun
14th Infantry, will proceed to Corinne
Utah, to identify a civilian there
in Confinement, under charge of
Complicity in the recent theft of
Arms at this post - Upon completion
of which duty he will rejoin his
proper station -
    By order of Captain Tobey
   "Sgd" Wm W. McCammon
     1st Lt & Adjt 14th Inftry
       Post Adjt.

Headquarters Camp Douglas U.T.
February 4th 1878.

Special Orders }
No 14   } &c

I At the request of the Quartermaster General U.S.A. and in obedience to orders from Headquarters Dept of the Platte January 23rd 1878, a Board of Survey to consist of
Captain Thomas F. Tobey,
1st Lieut John Murphy,   } 14 Infty
2nd Lieut Fred S. Calhoun,
will meet at this Post at 10 o'clock A.M. the 5th inst, for the following purpose, viz: To investigate and report upon 1800 lbs Hay and 2 Stocks & Dies, erroneously dropped from the returns of the late Capt E.B. Carling A.Q.M. U.S.A. at Camp Douglas U.T. for the 2nd Quarter 1874 which remain unaccounted for, also to investigate and report upon a balance of 171,233 lbs Hay and 3 Slates remaining on hand at Camp Douglas U.T. at close of 3rd Quarter 1874, for which the said Capt Carling was responsible, and which he carried forward as on hand on his returns for the 4th Quarter 1874, and 1st & 2nd Quarters 1875 as Post Quartermaster at Fort Saunders W.T.

By order of Colonel Smith
          Wm W. McCammon
          1st Lt r adjt 14th Infty & Post

Headquarters Camp Douglas U.T.
February 4th 1878

Special Orders
No 14 -   Ex
I At the request of the Quartermaster
General U.S.A. and in obedience to order
from Headquarters Dept of the Platte
January 23rd 1878, a Board of Survey
to Consist of
Captain Thomas F. Tobey
1st Lieut John Murphy   14th Infantry
2nd Lieut Fred S. Calhoun
will meet at this post at 10 O'Clock
A.M. the 5th inst. , for the following
purpose, Viz: To investigate and
report upon 1800 lbs Hay and
2 Stocks & Dies[?], erroneously dropped
from the returns of the late Capt
E.B. Carling A.Q.M. U.S.A. at Camp
Douglas U.T. for the 2nd Quarter 1874
which remains unaccounted for,
Also to investigate and report
upon a balance of 1712133 lbs
Hay and 3 slates remaining on
hand at Camp Douglas U.T. At
Close of 3rd Quarter 1874, for which
[crossed out] said Capt Cailing was responsible,
and which he carried forward
as on hand on his returns
for the 4th quarter 1874, and
1st & 2nd Quarter 1875 as Post
Quartermaster at Fort Saunders W.T.
            By order of Colonel Smith
            sgd Wm W. McCammon
            1st Lt Adjt 14th Inftry & Post

Hd Qrs Camp Douglas U.T.
January 26th 1878.

Special Orders
No 9

II A Board of Survey to consist of
Maj. B.A. Clements Surgeon U.S.A
1st Lieut John Murphy 14th Infty
2nd Lieut F. G. Calhoun 14th Infty
will assemble at this post at
once to receive a lot of Hay and
Straw to be delivered under Contract
at this Post –

The Board will keep an
account of the amount received,
and see that it is delivered in
accordance with the terms of
Contract – rendering report in
duplicate at this Office –

By Order of Colonel Smith
"Qys" Wm W. McCammon
1st Lt & Adjt 14th Infty & Actg

HdQrs Camp Douglas U.T.
January 23rd 1878.

Special Orders
No 9

II A Board of Survey to Consist of
Maj B.A. Clements Surgeon U.S. A.
1st Lieut John Murphy 14th Inftry
2nd Lieut FS. Calhoun 14th Inftry
will assemble at this post at
once to receive a lot of Hay and
Straw to be delivered under contract
at this post.

  The Board will keep an
account of the amount received ,
and see that it is delivered in
accordance with the terms of
Contract - rendering report in
duplicate of this Office -

By order of Colonel Smith
"sgd"   Wm W. McCammon
1st Lt & Adjt 14th Inftry & Post

Hd Qrs Camp Douglas U.T.
March 6" 1878.

Special Orders }
No 27

A Garrison Court Martial to Consist of
Surgeon B. A. Clements U.S.A
1 Lt Frank Taylor   14" Infty
2 Lt Fred S. Calhoun   14" Infty
is hereby appointed to Convene at
this post at 10 Oclock AM. the 9"
inst or as soon thereafter as practicable, for the trial of such Prisoners as may be properly brought
before it.
By order of Captain Burke
"Sig" Wm W McCammon
1 Lt & Adjt 14" Infty
Post Adjt.

Hd. Qrs Camp Douglas U.T.
March 23" 1878.

Special Orders }
No 35

I A Board of Survey to Consist of
Capt Thomas F. Tobey   14" Infty
1 Lt Frank Taylor   "   "
2d Lt F. Calhoun   "   "
is hereby Appointed to Convene at this post
at 10. Ock AM. this inst., or as soon thereafter
as practicable, for the following purpose.
To investigate and fix the responsibility for
shortage of Coal received from Captain J.V. Furey
A.Q.M. U.S.A. during the months of December 1877
and January 1878.
By order of Major M. Bryant
"Sig" Wm. W. McCammon
1 Lt & Adjt 14" Infty & Post.

Hd Qrs Camp Douglas U.T.
March 6th 1878

Special Orders
No 27
A Garrison Court Martial to Consist of
Surgeon B.A. Clements U.S.A.
1st Lt Frank Taylor 14th Inftry
2nd Lt Fred S. Calhoun 14th Inftry
is hereby appointed to convene at
this post at 10 O'Clock A.M. The 7th
inst. or as soon thereafter as practicable, for the trial of such prisoners as may be properly brought
before it.

By order of Captain Burke
"sgd" Wm W McCammon
1st Lt & Adjt 14th Inftry
Post Adjt.

Hd.Qrs Camp Douglas U.T.
March 23rd 1878

Special Orders
No 35
II A Board of Survey to Consist of
Capt Thomas F. Tobey 14th Inftry
1st Lt Frank Taylor 14th Inftry
2nd Lt F.S. Calhoun 14th Inftry
is hereby appointed to Convene at this post
at 10 Ock A.M. this inst., or as soon thereafter
as practicable, for the following purpose.
To investigate and fix the responsibility for
shortage of Coal received from Captain J.V. Furey
A.Q.M. U.S.A. during the months of December 1877
and January 1878.

By order of Major W. B ryant
"sgd" Wm W M. McCammon
1st Lt & Adjt 14th Inftry & Post

54

Move 5" Cos F, G, & I orders to Fort Hall I.T.

Hd Qrs Camp Douglas U.T. June 5" 1878

Special Orders
No 68

In obedience to telegraphic orders dated Hd Qrs Dept of the Platte 5" inst. Co G (Krause) Co F (Tobey) Co I (Taylor) 14" Infty will move at once to Fort Hall Idaho, prepared for field Service

By Order of Colonel Smith
"Sg" Wm W McCammon
1st Lt & Adj't 14" Infy & Post.

Hd Qrs Dept of the Platte
Omaha Bks Neb Novbr 2" 1878.

Special Orders
No 101 } Ext

On the arrival of recruits for Co A. 14" Inf at the Fort Hall Agency, Cos "F" "G" & "I" 14" Infy will proceed to Camp Douglas, U.T. (under Major Montgomery, Myauk 14" Inf) the Cmdg Officer of which post will order one of these Companies to proceed to Fort Cameron, Utah, and there take post, Major Myauk with the remaining Cos of his Command will report for duty at Camp Douglas.

By Command of Brig Genl Crook
"Sgs" Robert Williams
a a g.

[illegible] 5th Cos F G & I ordered to Fort Hall I.T.
   HdQrs Camp Douglas U.T. June 5th 1878
Special Orders
No 68
In obedience to telegraphic orders dated
HdQrs Dept of the Platte 5th inst. Co G. (Krause)
Co F. (Tobey) 14th Inftry will
move at once to Fort Hall Idaho, prepared
for field service
    By order of Colonel Smith
    "sgd" Wm W McCammon
     1st Lt &Adjt 14th Inftry & Post

   Hd Qrs Dept of the Platte
   Omaha Bks Neb Novbr 2nd 1878
Special Orders
No 101 - Ex
On the arrival of recruits for Co A 14th Inf
at the Fort Hall Agency, Co's "F," "G" &"I" 14th
Inftry # will proceed to Camp Douglas, U.T. (#under
Major Montgomery Bryant 14th Inf) the Com'dg
Officer of which post will order one of these
Companies to proceed to Fort Cameron, Utah,
and there take post, Major Bryant with the remaining
Co's of his Command will report for duty
at Camp Douglas.
   By Command of Brig Gen Crook
   "sgd" Robert Williams
    A.A.S.

Headquarters Fort Hall Idaho
June 16" 1878.

Special Orders
No 47

III Leave of Absence for Seven (7) days is granted 2nd Lt Fred G. Calhoun 14" Infty.

By order of Major Bryant
"Sgd" Thos B. Briggs
1st Lt 14" Infty
Post Adjt.

Headquarters Battalion 5" Cavalry
Ross Fork Idaho
Oct 19" 1878.

Orders.
Lieuts Calhoun & Kimball 14" Infantry, are hereby relieved from duty with Companies H & I 5" Cavalry, and will report to the Comdg Officer Fort Hall Agency.

In relieving these Officers the Battalion Commander is pleased to express his thanks for the cheerful and prompt manner in which they have performed all duties required of them.

Lieut Calhoun in Command of "H" Co 5" Cav at Ross Fork, Marched with great alacrity to Eagle Rock Bridge September 3d, in response to summons of the undersigned.

"Sgd" J M Hamilton
Captain 5" Cavalry
Commanding.

Headquarters Fort Hall Idaho
June 16th 1878

Special Orders
No 47
III Leave of Absence for seven (7)
days is granted 2nd Lt Fred S. Calhoun
14th Inftry

By order of Major Bryant
"sgd" Thos B Briggs
1st Lt 14th Inftry
Post Adjt

Headquarters Battalion 5th Cavalry
Russ[?] Fork Idaho.
Oct 19 1878

Orders.
   Lieuts Calhoun & Kimball 14th Infantry
are hereby relieved from duty with Companies
H & I. 5th Cavalry, and will report to the Com'dg
Officer Fort Hall Agency.
   In relieving these officers the Battalion
Commander is pleased to express his thanks
for the cheerful and prompt manner in which
they have performed all duties required of them.
   Lieut Calhoun in Command of "H" Co 5th Cav
at Ross Fork, marched with great alacrity
to Eagle Rock Bridge September 3d in response to
summons of the undersigned.

"sgd" J.M. Hamilton
Captain 5th Cavalry
Commanding.

Omaha Barracks Neb
Nov 10" 1878

Second Lieut Calhoun
14" Infty
Thro Comdg Officer
Camp Douglas Utah.

You are detailed as member on General Court Martial at Fort Fred Steele, to meet on the thirteenth inst.

By Command of Genl Crook
"Sgd" R. Williams
Asst Adjt General

Headquarters Dept of the Platte
Fort Omaha Neb January 18. 1879.

Special Orders }
No. 7   } Ex

2. Leave of Absence for One Month, with permission to apply for an extension of two months, is granted 2nd Lieut F. S. Calhoun, 14" Infty, (Fort Douglas, U. T.)

By Command of Brig Genl Crook
"Sgd" Robert Williams
Asst Adjt General.

Omaha Barracks Neb
Nov 10th 1878
Second Lieut Calhoun
14th Inftry
[illegible] Com'dg Officer
Camp Douglas Utah.

You are detailed as member on
General Court Martial at Fort Fred
Steele, to meet on the thirteenth inst. .
By Command of Genl Crook
"sgd" R Williams
Asst Adjt General

Headquarters Dept of the Platte
Fort Omaha Neb January 18. 1879
Special Orders
No 7 -   Ex/

2. Leave of Absence for One Month
with permission to apply for an
extension of two months, is granted
2nd Lieut F.S. Calhoun, 14th Inftry, (Fort
Douglas, U.T.)
By Command of Brig Gen Crook
"sgd"   Robert Williams
Asst Adjt General

Headquarters of the Army
Adjutant Gen'ls Office
Washington, February 14, 1879

Special Orders } Ex
No 34

5. The leave of absence granted 2d Lieutenant F. S. Calhoun 14th Infty, in Special Orders No 7, January 18, 1879, from Headquarters Dept of the Platte, is extended two months—

By Command of General Sherman
E. D. Townsend
Adjutant Gen'l

"O" R. C. Drum
Ass't Adj't Gen'l

Headquarters Fort Douglas U.T.
October 12" 1879.

Special Orders
No 133

V Lieut F. S. Calhoun 14" Infty, with Sergeant Clare Co K 14" Infty and one Private from Co F 14" Infty will proceed with three (3) pack animals, two (2) horses, and with necessary supplies, to explore the region about the head of City Creek Canyon— With a view to ascertaining and reporting upon the character and quantity of timber growing in that vicinity.

By order of Colonel Smith
"O" Wm W. McCammon
1st Lt & Adj'n 14" Inf'y & Post

> Headquarters of the Army
> Adjutant Gen'ls Office
> Washington February 14. 1879.

Special Orders
No 37 -   Ex
5. The leave of Absence granted 2d Lieutenant
F.S. Calhoun 14th Inftry, in Special Orders No 7, January
18, 1879, from Headquarters Dept of the Platte
is extended two months.
> By Command of General Sherman
> E.D. Townsend
> Adjutant Genl

"sgd" R.C. [illegible]
Asst Adjt Genl

> Headquarters Fort Douglas UT.
> October 12th 1879

Special Orders
No 133
II Lieut F.S. Calhoun 14th Inftry, with Sergeant
Clare C H 14th Inftry and one Private from Co
F. 14th Inftry will proceed with three (3) Pack
animals, two (2) Horses, and with
necessary supplies, to explore the
region about the head of City Creek
Canyon - With a view to ascertaining
and reporting upon the character and
quantity of [illegible] growing in that
vicinity.
> By order of Colonel Smith
> "sgd" Wm W. McCammon
> 1st Lt & Adjt 14th Inf & Post

Hdqrs Dist Black Hills
Camp Robinson Neb March 29, 1877

Special Orders } 2nd Lt F.G. Calhoun will accompany
No 3   } 1st Lt W.P. Clark 2nd Cavalry to
Spotted Tail Agency, Neb.

By order Col R.S. McKenzie
Sgd Jos H Dorst
1st Lt 4" Cav. a.a.a.genl

Omitted from Page 35.

Headquarters Fort Douglas U.T.
August 1" 1879.

Special Orders } 2d Lt F.G. Calhoun 14" Infty
No 100   } will proceed by the 3.40 train this
P.M. to Ogden Utah to receive and
conduct to this post a detachment
of recruits for 14" Infty.
 The Post Quartermaster will furnish
Transportation to Ogden for Lieut Calhoun
and Transportation for him and four
(4) Enlisted men returning

By order of Colonel Smith
Sgd Wm W. McCammon
1st Lt & Adjt 14" Inf & Post

Headqrs Dist Black Hills
Camp Robinson Neb March 29 1877

Official Orders
No 3   2nd Lt F.S. Calhoun will accompany
                1st Lt W.P. Clark 2nd Cavalry to
        Spotted Tail Agency, Neb.
                    By order Col R.S. Mackenzie
Omitted from   "sgd" Jos H Dorst
Page 35   2nd Lt 4th Cav AAA General

Headquarters Fort Douglas U.T.
August 1st 1879

Special Orders
No 100   2d Lt F.S. Calhoun 14th Inftry
will proceed by the 3:40 train this
P.M. to Ogden Utah to receive and
conduct to this post a detachment
of recruits for 14th Inftry.
    The Post Quartermaster will furnish
transportation to Ogden for Lieut Calhoun
and transportation for him and four
(4) Enlisted men returning

            By order of Colonel Smith
            sgd Wm W. McCannon
                1st Lt & Adjt 14th Inf & Post

FORT D. A. RUSSELL,
December 2, 1879.

GENERAL JOHN E. SMITH,

Commanding Fort Douglas:

My Dear General:—I take this, the first opportunity, since my return from the field, to express to you, their commanding officer, my high appreciation of the services of the officers and men of the Battalion of the 14th Infantry, lately in my command on the Ute Expedition; I always found that command cheerful, willing and anxious under the trying circumstances of the march and bivouac in the field, which after all, is the severest of a soldier's experience in campaign. I was only sorry that they were deprived of the chances of gaining the glory in a fight, which I know they would have achieved, if circumstances beyond the control of the military authorities had not prevented.

The Company Officers of the Battalion, without exception, are as perfect gentlemen and thorough soldiers, as it has ever been my good fortune to meet, and I count it a great gain to myself to have been associated with them as I was for a time on the expedition.

May I ask you, when opportunity occurs, to convey to the Battalion, just expressions of my appreciation of the soldierly qualities of both men and officers.

I am, my dear General, with great respect,

Faithfully your friend,

W. MERRITT, Colonel 5th Cavalry,

Brevet Major-General, U. S. A.

HEADQUARTERS FOURTEENTH U. S. INFANTRY,
Fort Douglas, Utah, December 5, 1879.

In publishing the above, the Colonel commanding the regiment desires to express the gratification it affords him to have the conduct of the Fourteenth Infantry Battalion so highly commended. The appreciation of the soldierly qualities of the officers and men, by so able a soldier as General Merritt, re-assures the Regimental Commander that when called upon, the qualities commended, will render them equal to any emergency.

JOHN E. SMITH, Colonel 14th Infantry,

Commanding Regiment.

Official:

First Lieut. and Regimental Adjutant.

Custer's Other Brother-in-Law

Fort D. A. Russell
December 2, 1879

General John E. Smith,

                      Commanding Fort Douglas:

My Dear General: - I take this, the first opportunity, since my return from the field, to express to you, their commanding officer, my high appreciation of the services of the officers and men of the Battalion of the 14th Infantry, lately in my command on the Ute Expedition; I always found that command cheerful, willing and anxious under the trying circumstances of the march and bivouac in the field, which after all, is the severest of a soldier's experience in campaign. I was only sorry that they were deprived of the chances of gaining the glory in a fight which I know they would have achieved, if circumstances beyond the control of military authorities had not prevented.

The Company Officers of the Battalion, without exception, are as perfect gentlemen and thorough soldiers, as it has ever been my good fortune to meet, and I count it a great gain to myself to have been associated with them as I was for a time on the expedition. May I ask you, when opportunity occurs, to convey to the Battalion, just expressions of my appreciation of the soldierly qualities of both men and officers.

      I am, my dear General, with great respect,
           Faithfully your friend,
                  W. Merritt, Colonel 5th Cavalry,
                        Brevet Major-General, U.S.A.

Headquarters Fourteenth U. S. Infantry,
Fort Douglas, Utah, December 5, 1879.

In publishing the above, Colonel commanding the regiment desires to express the gratification it affords him to have the conduct of the Fourteenth Infantry Battalion so highly commended. The appreciation of the soldierly qualities of the officers and men, by so able a soldier as General Merritt, re-assures the Regimental Commander that when called upon, the qualities commended, will render them equal to any emergency.

                John E. Smith, Colonel 14th Infantry,
Official: Commanding Regiment.
      [signature of 1LT William W. McCammon],
      First Lieut. and Regimental Adjutant.
[In red ink:] 2d Lieut F.S. Calhoun, 14th Infy.

Headquarters Fort Douglas U.T.
March 4th 1876

Special Orders
No 31

2nd Lt F. S. Calhoun 14th Infantry will proceed on Sunday Morning next En route to Fort Cameron U.T. in Charge of the Recruits for Companies of the 14th Infantry stationed at that Post.

The detachment will be furnished with rations to include the 10th inst. Having turned over the recruits at Fort Cameron Lieut Calhoun will rejoin his proper station.

The A. Q. M. Dept will furnish the necessary transportation to terminus of the U. P. R.R.

By Order of Col Geo E. Smith

W. H. W. C. Cameron
1st Lt & Adjt 14th Inf'y & Post

Headquarters Fort Douglas U.T.
March 4th 1880

Special Orders
No 31

2nd Lt F.S. Calhoun 14th Infantry will proceed on Sunday morning next En-route to Fort Cameron U.T. in charge of the Recruits for Companies of the 14th Infantry stationed at that post.

The detachment will be provided with rations to include the 10th [illegible]. Having turned over the recruits at Fort Cameron Lieut Calhoun will rejoin his proper station. The Q.M. Dept will furnish the necessary transportation to terminus of the U.S. R.R.

By order of Col Jm. E. Smith
sgd Wm W McCammon
1st Lt Adjt 14th Inf & Post

Headquarters Fort Douglas U.T.
April 8" 1880.

Special Orders (
No 51)

II Under Telegraphic authority dated Headquarters Dept. of the Platte, Fort Omaha Neb April 6" 1880 — Lieut Fred S. Calhoun 14" Infantry, is hereby authorized to anticipate the sick leave he has applied for, and is hereby relieved from Post duty.

By order of Colonel Smith
"Sgd" Wm W. McE. Cannon
R.Q. & adj't 14" Infty & Post.

Left Fort Douglas En-route to Chicago April 10th 1880.

Headquarters Department of the Platte
Fort Omaha Nebraska, April 16" 1880.

Special Orders (
No 34)

1. Leave of Absence for one month, on Surgeon's certificate of disability, is granted Second Lieutenant Frederick S. Calhoun, 14" Infantry, (Fort Douglas U.T.)

By command of Brig Gen Crook
"Sgd" Robert Williams
Assistant Adj't General

This leave was extended one month at a time on ~~Sur~~ Certificates of Disability until September 9" 1880. Reported for duty at Fort Douglas U.T. Sept 8" 1880.

Headquarters Fort Douglas U.T.
April 8th 1880

Special Orders
No 51

II Under Telegraphic authority dated Headquarters Dept. of the Platte. Fort Omaha Neb April 6th 1880 - Lieut Fred S. Calhoun 14th Infantry is hereby authorized to anticipate the sick leave he has applied for and is hereby relieved from Post duty.

By order of Colonel Smith
"sgd" Wm W. McCammon
1LT & adjt 14th Inftry & Post.

Left Fort Dougland En-route to Chicago April 10th 1880.

Headquarters Department of the Platte
Fort Omaha Nebraska, April 16th 1880

Special Orders
No 34

I. Leave of Absence for One month, on Surgeons certificate of disability, is granted Second Lieutenant Frederic S. Calhoun, 14th Infantry, (Fort Douglas U.T.)

By Command of Brig Gen Crook
sgd"   Robert Williams
assistant Adjt General

This leave was extended one month at a time Certificates & Disability until September 9th 1880

Reported for duty at Fort Douglas U.T. April 8th 1880 [in the journal, the entry, including the word 'April' were written in black ink. The word April was later overwritten in green, and replaced with 'Sept'.]

Headqurters Fort Douglas U.T.
January 31" 1881.

Special Orders }
No 21   } &c

V Lieut F.S. Calhoun 14" Infty, will
proceed on Wednesday Feby 2nd 1881
Enroute to Fort Hall I.T. in charge
of Eight (8) Recruits for Company
"A". 14" Infantry. The recruits will
be provided with two (2) days
Cooked rations, and the
Quartermasters Dept will furnish
the necessary transportation.

VI Having turned over the recruits
which he is ordered to conduct
to Fort Hall I.T. Lieut F.S. Calhoun
14" Infty, will rejoin his proper Station.

By order of Col Smith
"Sgd" Wm W. McCammon
1" Lt & Adjt 14" Infty & Post

Left Douglas Feby 2nd Arrived at
Fort Hall Feby 3rd. Returning left
Fort Hall Feby 4" arrived at Fort
Douglas Feby 5.
Reported for Duty Feby 6th

Detailed for Officer of Guard and
served tour Feby 7th

Did not give me time to take a
Bath or blow my nose before
putting me on garrison guard duty.

Headquarters Fort Douglas U.T.
January 31st 1881

Special Orders
No 21 -   Ex
V   Lieut F. S. Calhoun 14th Inftry will
proceed on Wednesday Febry 2nd 1881
Enroute to Fort Hall I.T. in charge
of Eight (8) Recruits for Company
A. 14th Infantry. The recruits will
be provided with two (2) days
Cooked rations, and the
Quartermasters Dept will furnish
the necessary transportation.
VI   Having turned over the recruits
which he is ordered to conduct
to Fort Hall I.T. Lieut F.S. Calhoun
14th Inftry, will rejoin his proper station.

By order of Col Smith
"sgd" Wm W.McCammon
1st Lt & Adjt 14th Inftry & Post

[Personal note in black ink below:]

Left Douglas Febry 2nd Arrived at
Fort Hall Febry 3rd. Returning left
Fort Hall Febry 4th Arrived at Fort
Douglas Febry 5.
<u>Reported for Duty Febry 6th</u>
<u>Detailed for Officer of Guard and served tour Febry 7th</u>
Did not give me time to take a
Bath or blow my nose before
putting me on garrison guard duty.

Headquarters Fort Douglas Utah
February 17th 1881

Lieut Fred S. Calhoun 14" Infty
  Fort Douglas Utah
(thro Lieut Austin 14" Infty)

Sir:
The Post Commander directs that the order of yesterday, placing Lieut Austin 14" Infty in charge of the Quartermasters and Sub Departments at this Post - during the absence of Lieut Patterson, be revoked - and that Lieut Calhoun have supervision of the Depts referred to -

Very Respectfully
Your Obt Servant
"sgd" Wm W McCammon
  1st Lt & Adjt
  14" Infty & Post.

———

Headquarters Fort Douglas Ut.
Feby 22nd 1881.

Special Orders) Et
   No 35.   )

I. In obedience to telegraphic orders dated Headquarters Dept of the Platte - this instant, Lieut Fred S. Calhoun 14" Infty will proceed without delay to Fort Omaha Nebraska, reporting on arrival there to Brigadier General Crook U.S.A. as witness in the case of Captain Charles B. Western, 14" Infty - Upon completion of which duty, he will rejoin his proper station.
            By order of Major Bryant

Headquarters Fort Douglas Utah
February 17th 1881
Lieut Fred S. Calhoun 14th Inft
Fort Douglas Utah
(thro Lieut Austin 14th Inftry)

Sir. The Post Commander directs that the Orders of yesterday, placing Lieut Austin 14th Inftry in charge of the Quartermasters and Sub Departments at this post - during the absence of Lieut Patterson, be revoked, and that Lieut Calhoun have supervision of the Depts referred to.
    Very Respectfully
      Your Off Servant
"sgd" Wm W.McCammon
  1st Lt & Adjt
    14th Inftry & Post.

Headquarters Fort Douglas U.T.
Febry 22nd 1881

Special Orders
No 35 -   Ex
I In obedience to telegraphic orders dated Headquarters Dept of the Platte - this instant Lieut Fred S. Calhoun 14th Inftry will proceed without delay to Fort Omaha Nebraska, reporting on arrival there to Brigadier General Crook U.S.A. As witness in the case of Captain Charles B. Western, 14th Inftry. Upon completion of which duty, he will rejoin his proper station
        By order of Major Bryant

64

(Signed) Wm W McCammon
1st Lieut & adjt. 14th Infty
Post Adjutant.

Sent On Feby 23rd
Returned March 5th

HEADQUARTERS FOURTEENTH U. S. INFANTRY,
Fort Douglas, Utah, May 20, 1881.

GENERAL ORDERS,
No. 4.

The President of the United States having retired the undersigned to date 19th inst., severs his connection with the 14th Regiment U. S. Infantry.

The ties formed by an association of over eleven years, cannot be severed without sincere regret on the part of your late Regimental Commander, but it only remains for him to acknowledge the courtesy of the officers who have uniformly given him their support in the discharge of his duties, in which he has always had in view the interests of the service, together with the welfare of the whole command. The few years that may be spared him, will be spent by your late Regimental Commander in friendly remembrance of you all, and to the officers and enlisted men of the Regiment, he tenders his earnest wishes for their welfare. Trusting that you will remember him kindly, relinquishing command of the 14th Infantry, he bids you all, farewell.

Colonel 14th Infantry.
Bvt Maj Genl USA

HEADQUARTERS FOURTEENTH U. S. INFANTRY,
Fort Douglas, Utah, July 30, 1881.

GENERAL ORDERS,
No. 5.

In accordance with Par. 6, of Special Orders, No. 118, Headquarters of the Army, A. G. O., current series, the undersigned hereby assumes command of this regiment.

[S'g'd]   L. C. HUNT,
            Colonel 14th Infantry.

OFFICIAL:

First Lieut. and Adjutant 14th Infantry.

(signed) Wm W McCammon
1st Lieut & Adjt 14th Inftry
Post Adjutant.

Left on Feby 23rd
Returned March 3rd

HEADQUARTERS, 14th U. S. INFANTRY
Fort Douglas. Utah, May 20, 1881.

General Orders:
No. 4

The President of the United States having retired the undersigned to date 19th Inst., severs his connection with the 14th Regiment of the U.S. Infantry

The ties formed by an association of over eleven years, cannot be severed without sincere regret on the part of your late-Regimental Commander, but it only remains for him to acknowledge the courtesy of the officers who have uniformly given him their support in the discharge of his duties, in which he has always had in view the interests of the service, together with the welfare of the whole command. The few years that may be spared him, will be spent by your late Regimental Commander in friendly remembrance of you all, and to the officers and enlisted men of the Regiment, he tenders his earnest wishes for their welfare. Trusting that you will remember him kindly, relinquishing command of the 14th infantry, he bids you all, farewell.

[signed]   Jm p Smith
Bvt Maj Genl USA

HEADQUARTERS, 14th U. S. INFANTRY
Fort Douglas. Utah, May 20, 1881.

General Orders:
No. 5

IIn accordance with Par. 6, of Special Orders, No. 118, Headquarters of the Army, A.G.O., current series, the undersigned hereby assumes command of this regiment
[S'g'd]   L. C. Hunt
Colonel 14th Infantry

OFFICIAL:

William W. McCammon
1st Lieut. and Adjutant 14th Infantry

Hd Qrs Dept of the Platte
Fort Omaha Neb. June 28. 1881

Special Orders } Ex
No 59

I. A General Court Martial is hereby appointed to meet at Fort Hall, Idaho, on the 7th day of July 1881, or as soon thereafter, as practicable, for the trial of such prisoners as may be brought before it.

Detail for the Court:
1. Captain A. H. Bainbridge 14" Inf
2. 1st Lt Chas H Warrens "
3. 1st Lt Frank Taylor "
4. 2nd Lt R. T. Yeatman "
5. " F. S. Calhoun "
6. " W. B. Reynolds "

1st Lieut J. E. Quentin, 14" Inf Judge advocate —

No other officers than those named can be assembled without manifest injury to the service.

By order of Brig Gen Crook
Comdg Dept of the Platte
"sgd" Robert Williams
A A G

Head Quarters Dept of Platte
Fort Omaha Neb.
July 19" 1881

Special Orders } Ex
No 67

I. A General Court Martial is hereby appointed to meet at Fort Douglas, Utah, on the 25th day of July 1881, or as soon thereafter as practicable, for the trial of such prisoners as may be

HdQrs Dept of the Platte
Fort Omaha Neb. June 28, 1881

Special Orders
No 59 -   Ex
I. A general Court Martial is hereby appointed to meet at Fort Hall, Idaho, on the 7th day of July 1881, or as soon thereafter as practicable, for the trial of such prisoners as may be brought before it.

Detail for the Court:
1 Captain A.H. Bainbridge   14th Inf
2 1 Lt Chas H Warrens   "
3 1st Lt Frank Taylor   "
4 2nd Lt R.T. Yeatman   "
5 2nd Lt F.S. Calhoun   "
6 2nd Lt W.B Reynolds   "

1st Lieut J.E. Quentin, 14th Inf Judge Advocate
No other officers than these named can be assembled without manifest injury to the service.

By order of Brig Gen Crook
Com'dg Dept of the Platte
"sgd" Robert Williams
A.A.G.

Head Quarters Dept of Platte
Fort Omaha Neb.
July 19th 1881

Special Orders
No 67 -   Ex
I. A general court martial is hereby appointed to meet at Fort Douglas, Utah. on the 25th day of July 1881, or as soon thereafter as practicable, for the trial of such prisoners as may be

brought before it

Detail for the Court.

1. Major Saml. M. Horton Med Dept USA
2. Capt Gilbert S. Carpenter 14" Inf
3. "   Saml. McConihe   "
4. "   Chas B. Western   "
5. 1st Lt Chas H. Warrens   "
6. "   Albert Austin   "
7. "   Frank Taylor   "
8. 2nd Lt Frederic S. Calhoun "
9. "   Stephen J. Mulhall   "

Capt Thos F. Tobey 14" Infty
Judge Advocate –

No other officers than those named can be assembled, without manifest injury to the service.

+   +   +   +

By order of Brig Genl Crook
Comdg the Dept of the Platte
"Sgd" Robert Williams
asst Adjt Genl

    brought before it
    Detail for the Court.
1 Major Sam'l M. Horton Med Dept U.S.A.
2 Capt Gilbert S. Carpenter   14th Inf
3 Capt Sam'l McCarribe   "
4 Capt Chas B. Western   "
5 1st Lt Chas H. Warrens   "
6 1st Lt Albert Austin   "
7 1st Lt Frank Taylor   "
8 2nd Lt Frederic S. Calhoun   "
9 2nd Lt Stephen J. [illegible]   "
Capt Thos F. Tobey 14th Inftry
Judge Advocate.
 No other officers that these named
can be assembled, without manifest
injury to the service
   X X X   X
  By order of the Brig Gen Crook
    Comdg the Dept of the Platte
     "sgd" Robert Williams
       Asst Adjt General

Battalion 14th Infantry
Cheyenne Wyo, Territory
August 9th 1881.

Special Orders }
No 2   } Ex

III In compliance with telegraphic instructions from Headquarters Dept of the Platte, dated Fort Omaha, Neb, August 9th 1881, 2nd Lieut F.F. Calhoun, 14th Infantry, and One Sergeant and three privates of Co "G." One Corporal and two privates of Co "F." three privates of Co "B." and two privates of Co "D." 14th Infty, will proceed to Fort Lyon Colorado, and take Station at that Post.

The Quartermaster Department will furnish the necessary transportation.

By Order of Lieut Col Douglas
"Sgd" J. E. Quentin
1st Lieut 14th Infty
Actg Adjt.

Headquarters Fort Lyon Colo
August 12th 1881

Orders }
No 72 }

I 2nd Lt F.F. Calhoun, 14th Infty, having reported for duty at these Headquarters, is hereby appointed a.a.q.m. — a.a.c.s. — a.o.o. and a.s.o. at this post.

Lieut Col D. Huston 6th Infantry will turn over all property and funds pertaining to the above mentioned offices, exchanging the usual invoices & receipts.

"Sgd" Dan Huston
Lieut Col 6th Infty
Bvt Col U.S.A. Comdg

Battalion 14th Infantry,
Cheyenne Wyo Territory
August 9th 1881

Special Orders
No 2 -   Ex
II In compliance with telegraphic instructions from Headquarters Dept of the Platte, dated Fort Omaha, Neb. August 9th 1881, 2nd Lieut. F.S. Calhoun 14th Infantry, and one Sergeant and three privates of Co "G" One Corporal and two privates of Co F. three privates of Co "H" and two privates of Co "D" 14th Inftry, will proceed to Fort Lyon Colorado, and take Station at that post.
             The Quartermaster Department will furnish the necessary transportation
         By order of Lieut. Col Douglas
              "sgd" J.E. Quentin,
                1st Lieut 14th Inftry
                   Actg Adjt

Headquarters Fort Lyon Colo
August 12 1881

Orders
No 72
     I   2nd Lt F.S. Calhoun 14th Inftry having reported for duty at these Headquarters, is hereby appointed A.A.Q.M. - A.A.C.S. - A.O.O. and A.S.O. at this post.
Lieut Col D. Huston 6th Infantry will turn over all property and funds pertaining to the above mentioned offices, exchanging the usual invoices & receipts.
          "sgd"   Dan Huston
             Lieut. Col 6th Infty
             Bvt Col U.S.A. Cmding

Headquarters Fort Lyon Colo
August 18" 1881

Orders }
No 75 }
☩ The Undersigned hereby
assumes Command of this post.

"By" Fred S. Calhoun
2nd Lieut 4" Infantry
Com'dg

Headquarters Dep't of the Mo
Asst Adjutant Gen'ls Office
Fort Leavenworth Kansas
October 15" 1881.

Special

Left Fort Lyon Oct 25" 1881
Arrived at Cantonment on Uncompahgre
Colorado Oct 30" 1881.

Headquarters Fort Lyon Colo
August 18th 1881

Orders No
75

I The Undersigned hereby
assumed Command of this post

"   sgd"   Fred S. Calhoun
2nd Lieut 4th Infantry
Com'dg

Headquarters Dept of the Mi.
Asst Adjutant Genls Office
Fort Leavenworth Kansas
October 12th 1881

Special
[Below the word special, Calhoun pinned a printed handout to the page. This has been included with this book on the next two pages. This printout is of Special Order No. 28, directing troop movements within the Division of the Missouri, including his own.]

Left Fort Lyon Oct 25th 1881
arrived at Cantonment on the Unconpahgre
Colorado Oct 30th 1881

HEADQUARTERS DEPARTMENT OF THE MISSOURI,
ASSISTANT ADJUTANT-GENERAL'S OFFICE,
Fort Leavenworth, Kansas, October 12, 1881.

SPECIAL ORDERS,
No. 208.

EXTRACT.

1. The following movements of troops and changes of stations are hereby ordered:

Three companies of the 13th Infantry, from Fort Lewis, Colorado, will proceed, under command of the Lieutenant-Colonel of the regiment, to Fort Craig, New Mexico, and report to Colonel L. P. *Bradley*, commanding the District of New Mexico, to relieve the same number of companies of the 15th Infantry, to be selected by the Colonel of that regiment, who will proceed with them to take post at Fort Lewis, Colorado, where he will establish the Headquarters of his regiment. When he reaches Fort Lewis the remaining companies of the 13th Infantry at that post will proceed to relieve the two companies of the 15th Infantry now on the line of the Atlantic and Pacific Railroad west of Albuquerque, which companies, when so relieved, will proceed to take post at Fort Lewis, Colorado.

The horses and equipments of the mounted company of the 13th Infantry, now at Fort Lewis, will be turned over to a company of the 15th Infantry at that post, to be selected by the Colonel, and the detachment at Pagosa, Colorado, will be relieved by one of the 15th Infantry, to be sent from Fort Lewis.

The five companies of the 23d Infantry now at the Cantonment on the Uncompahgre, are transferred to New Mexico, as follows:

Four companies to Fort Union, New Mexico, to which point the Headquarters of the regiment are transferred. The remaining company will proceed to Fort Craig, New Mexico, and report to the District Commander for assignment to a post. The Major of the regiment will also report to the District Commander for assignment.

Four additional companies of the 23d Infantry to be taken, as follows: Two from Fort Dodge, Kansas, one from Fort Wallace, Kansas, and one (G) from Fort Reno, I. T., will proceed without delay to Fort Craig, New Mexico, and report to the District Commander for assignment.

When these four companies of the 23d Infantry reach Fort Craig, New Mexico, three companies of the 15th Infantry will be sent under command of the Lieutenant-Colonel of the regiment, to take post at Fort Lyon, Colorado. The Major of the regiment will take post at Fort Lewis, Colorado. One company (B) will be sent to take post at Fort Garland, Colorado; the remaining company of the regiment (I) will be held at Santa Fé, New Mexico, until further orders.

One company of the 19th Infantry from Fort Leavenworth, Kansas, will proceed to Fort Dodge, Kansas, and take post there, relieving the two companies of the 23d Infantry, and will send a commissioned officer and small detachment to Fort Lyon, Colorado, relieving the detachment of the 14th Infantry now there, which, when so relieved, will proceed to join its command at the Uncompahgre, *via* Gunnison, Colorado, without delay.

The detachment of the 14th Infantry now at Fort Garland, Colorado, will remain there until relieved by the arrival of the company of the 15th Infantry ordered there, and will then proceed to join its command at the Uncompahgre.

2

The commanding officer of Fort Hays, Kansas, will send a commissioned officer and small detachment to take charge of Fort Wallace, Kansas, until further orders.

The battalion of the 14th Infantry under command of Lieutenant-Colonel *Henry Douglass*, will take post at the Cantonment on the Uncompahgre, and will constitute the garrison of that post.

One company of the 24th Infantry will be sent from the Cantonment Indian Territory to Fort Reno, I. T., to replace at that post the company of the 23d Infantry ordered sent from there.

The movements herein ordered will take place with the least practicable delay, and will be made under the direction of the Commanding Officer District of New Mexico and the several post commanders concerned.

The Chief Quartermaster of the Department will take the necessary measures to furnish transportation to carry out the provisions of this order, and for this purpose will take immediate steps to ascertain from the several post commanders the exact time when the transportation will be needed for their respective commands.

BY COMMAND OF BRIGADIER-GENERAL POPE:

E. R. PLATT,
Assistant Adjutant-General.

OFFICIAL:

E. R. Platt
Assistant Adjutant-General.

Headquarters Fort Lyon Colo
October 24" 1881

Orders }
No. 7 } II In compliance with S.O.
No. 208. Headquarters Department
of the Mo., the detachment of the
14" Infantry are relieved from duty
at this post and will proceed
to join its Command at the
Uncompahgre, Colorado, without delay.
The Quartermasters Department
will furnish the necessary transportation, and the Subsistence
Department, Commutation of rations
for the necessary number of
days at the usual rates in
advance, it being impracticable
to carry cooked rations.

Fred S. Calhoun
2nd Lieut 14" Infantry
Comdy Post

———

Headquarters Dept of the Missouri
Adjutant Generals Office.
Leavenworth, Kansas,
February 8" 1882.

Special Orders }
No. 30. } Ex.
4. Leave of Absence for One (1) Month,
with permission to apply for an extension
of One (1) month, is granted 2d Lieut
Frederick S. Calhoun, 14" Infty (Cantonment on Uncompahgre Co.
By order of Brig-Gen Pope

"Sgd" E. R. Platt
Major †Asst Adjt Genl U.S. a
Adjt General.

Headquarters Fort Lyon Colo
October 2nd 1881

Orders
 No 87   II In compliance with S.O
No 208 Headquarters Department
of the Mi, the detachment of the
14th Infantry are relieved from duty
at this post and will proceed t
o join its command at the
Uncompahgre, Colorado without delay.
  The quartermasters Department
will furnish the necessary transportation, and the Subsistence
department, [illegible] of rations
for the necessary number of
days at the usual rates in
advance, it being impracticable
to carry cooked rations.
             sgd   Fred S. Calhoun
                    2nd Lieut 14th Infantry
                    Com'dg Post

Headquarters Dept of the Missouri
Adjutant Generals Office
Leavenworth, Kansas,
February 8th 1882

Special Orders
No. 30 -   Ex
4. Leave of Absence for one (1) month
with permission to apply for an extension
of one (1) month, is granted 2d Lieut
Frederic S. Calhoun, 14th Inftry Cantonment on Uncompahgre Colo.
By order of Brig. Gen Poho[?]
        sgd" E.R. Platt Major & Asst Adj't Genl U.S.A.
                Adjt General.

Cantonment on Uncompahgre
March 1" 1882.

Orders
No 49

To enable 2d Lieut. Fred S. Calhoun 14" Infantry, to avail himself of the leave granted under S.O. #30 from Headquarters Department of Missouri, he is hereby relieved from duty at this post.

By order of Lt Col Douglass
"Sgd" J. E. Guentin
1" Lt 14" Infty
Post Adjutant.

———

Headquarters Mil. Div. of the Missouri
Chicago, Ills, March 13" 1882.

Special Orders
No. 28

1. The leave of absence granted 2d Lieutenant Frederic S. Calhoun, 14" Infantry, (Cantonment on Uncompahgre, Colorado,) by paragraph 4, S. O. No 30, dated Hd. Qrs. Dept. of Mo. February 8" 1882, is extended one month.

By Command of Lt General Sheridan
"Sgd" Robert Williams
Asst Adjt. General.

Cantonment on Uncompahgre
March 1st 1882

Orders
No 49
To enable 2d Lieut Fred S. Calhoun
14th Infantry, to avail himself of the
leave granted under S.O. #30 from
Headquarters Department of Missouri,
he is hereby relieved from duty at
this point
        By order of Lt Col Douglas
          "sgd" J.E. Quentin
            1st Lt 14th Infty
              Post Adjutant

Headquarters Mil. Div. of the Missouri
Chicago, Ills, March 3rd 1882

Special Orders
No. 28
1. The leave of absence granted 2d
Lieutenant Frederic S. Calhoun,
14th Infantry, (Cantonment on the Uncompahgre, Colorado.) by paragraph 4.
S.O. No 30, dated Hd.Qrs. Dept
of Mo. February 8th 1882, is extended
one month.
  By Command of Lt General Sheridan
        "sgd" Robert Williams
           Asst Adjt General.

Headquarters of the Army
Adjutant General's Office
Washington April 11. 1882.

Special Orders }
No. 83 } Ext.

5. The extension of leave of absence granted 2d Lieutenant Frederic S. Calhoun, 14th Infantry, in Special Orders, No. 28, March 13" 1882, Military Division of the Missouri, is further extended One Month.

By Command of General Sherman
 R.C. Drum
 Adjutant General.

Cantonment on Uncompahgre, Colorado,
October 15th 1882.

Orders }
No. 204 }

2nd Lieutenant Fred S. Calhoun 14" Infantry is hereby granted leave of absence for Seven days.

By order of Lt. Col. Douglass
 J.E. Quentin
 1st Lieut 4" Infty
 Post Adjutant.

Headquarters of the Army
Adjutant Generals Office
Washington April 11. 1882.

Special Orders
No. 83 -   Ex

5. The extension of leave of absence granted 2d Lieutenant Frederic S. Calhoun, 14th Infantry, in Special Orders, No. 28, March 13th 1882, Military Division of the Missouri, is further extended for one month.

By Command of General Sherman
"sgd" R.C. Down[?]
Adjutant General.

Cantonment On Uncompahgre, Colorado,
October 10th, 1882

Orders
No 204

2nd Lieutenant Fred S. Calhoun 14th Infantry is hereby granted leave of absence for seven days.

By order of Lt Col Douglas
"sgd" J.E. Quentin,
1st Lt 14th Inftry,
Post Adjutant.

Cantonment On Uncompahgre
May 31" 1883.

Orders }
No 105 }

Pursuant to Authority granted this day by telegram from Head-quarters Department Mo. Lieut. F. S. Calhoun with Corporal Mc Aleer will proceed to Gunnison Colo, and receive from the City Marshal three apprehended deserters, giving receipts for the same, and returning with the same to this Post. The Q. M. Dept will furnish the necessary transportation.

The A.C.S. will furnish Corporal Mc Aleer with One days travel ration —

By order of Lt Col Douglass
"Sgd" J. E. Quenten
1"Lt 14" Infty
Post Adjutant

Cantonment On Uncompahgre
May 31st 1883

Orders
No 105

Pursuant to Authority granted this day by telegram from Head quarters Department Mo. Lieut F.S. Calhoun with Corporal Mc Aleer will proceed to Garrison[?] Colo, and receive from the City Marshal three apprehended deserters, giving receipt for the same, and returning with the same to this Post. The Q.M. Dept will furnish the necessary transportation.

The A.C.S. will furnish Corporal McAleer with one days travel Ration.

By order of Lt Col Douglass
"sgd" J.E. Quentin
1st Lt 14th Infty
Post Adjutant

HEADQUARTERS DIVISION OF THE PACIFIC,
Presidio of San Francisco, Cal., September 11, 1884.

SPECIAL ORDERS
No. 92.

(Extract.)

1.—Leave of absence for two months is hereby granted Second Lieutenant *F. S. Calhoun*, 14th Infantry, Vancouver Barracks, W. T., with permission to go beyond the limits of this Division and to apply to the Adjutant General of the Army for an extension of two months.

* * * * *

By command of Major General POPE:

J. C. Kelton
Assistant Adjutant General.

SPECIAL ORDERS,
No. 244.

HEADQUARTERS OF THE ARMY,
ADJUTANT GENERAL'S OFFICE,
Washington, October 17, 1884.

Extract.

1. The leave of absence granted 2d Lieutenant *Frederic S. Calhoun*, 14th Infantry, in Special Orders, No. 92, September 11, 1884, Division of the Pacific, is extended two months.

* * * * *

BY COMMAND OF LIEUTENANT GENERAL SHERIDAN:

R. C. DRUM,
Adjutant General.

OFFICIAL:

Thomas Ward
Assistant Adjutant General.

Co. C 14" Infty, to Co. to which I was temporarily attached, left Vancouver Barracks W.T. Nov. 7 # 1885 # and arrived in Seattle at 3. A.M. Nov. 8" Remained in Seattle until Nov. 17" returned to Vancouver Bks —

This movement was caused by the Knights of Labor attempting to drive off the Chinese —

\# Per Orders No 269 – Vanc. Barracks W.T. Nov. 6" 1885.
\#\# Orders No —

HEADQUARTERS DIVISION OF THE: PACIFIC,
Presidio of San Francisco, Cal., September 11, 1884
Special Orders
No. 92.

(Extract.)

1.-Leave of absence for:: two months is hereby granted Second Lieutenant *F.* Calhoun, 14th Infantry, Vancouver Barracks. W. T., with permission to go beyond the limits of this Division and to apply to the Adjutant General of the Army for an extension of two months.

By command of Major General Pope:
Assistant Adjutant General

Special Orders, } HEADQUARTERS OF THE ARMY,
No. 244 ADJUTANT GENERALS OFFICE
Washington, October 17', 1884.
Extract.

I. The leave of absence granted 2d Lieutenant Frederic S. Calhoun, 14th Infantry, in Special Orders, No. *92,* September 11, 1884 Division of the Pacific, is extended two months.

BY COMMAND OF LIEUTENANT GENERAL SHERIDAN:
R. C. DRUM,
Adjutant General.

[The following entry was written in ordinary pencil, one of Calhoun's few personal notes in this book.]

Co. C 14th Infty, to Co, to which I was
temporarily attached, left Vancouver
Barracks W.T. Nov 7th 1885 and arrived
in Seattle at 3 A.M. Nov. 8th. Remained
in Seattle until Nov 17 returned to
Vancouver Bks. –
This movement was caused by the Knights
of Labor attempting to drive off the
Chinese.
Per orders Uo 269. Vanc. Barracks W.T. Nov. 6th 1885.
Orders No.

74

> Left Vancouver Bks with Cat in Exp'd'n
> to Seattle W.T. per order No. 29 Vancouver
> Barracks W.T. Feby 8" 86 — on Feby 9th
> 1886. arrived at Seattle Feby 10th 1886
> Returned to Vanc. Bks Feby 20" '86
>
> Knights of labor riots against Chinese
> was the cause of this movement of troops.

HEADQUARTERS DEPARTMENT OF THE COLUMBIA,
Vancouver Barracks, W. T., March 15, 1886.

SPECIAL ORDERS }
No. 43. }

7.—A General Court-Martial is appointed to meet at Fort Walla Walla on Monday, the 22d instant, at 10 o'clock, A. M., or as soon thereafter as practicable, for the trial of such prisoners as may be properly brought before it.

DETAIL FOR THE COURT:
1. Lieutenant-Colonel *John Green*, 2d Cavalry.
2. Major *Thomas McGregor*, 2d Cavalry.
3. Captain *James N. Wheelan*, 2d Cavalry.
4. Captain *Eli L. Huggins*, 2d Cavalry.
5. Captain *Samuel T. Hamilton*, 2d Cavalry.
6. 1st Lieutenant *John Murphy*, 14th Infantry.
7. 1st Lieutenant *Charles B. Schofield*, 2d Cavalry.
8. 1st Lieutenant *Benjamin Munday*, Assistant Surgeon.
9. 2d Lieutenant *Frederic S. Calhoun*, 14th Infantry.
10. 2d Lieutenant *Thomas J. Lewis*, 2d Cavalry.
11. 2d Lieutenant *Frederick D. Holton*, 2d Cavalry.
12. 2d Lieutenant *Alfred Hasbrouck, jr.*, 14th Infantry.
13. 2d Lieutenant *Henry C. Cabell, jr.*, 14th Infantry.

1st Lieutenant *Frederick W. Kingsbury*, 2d Cavalry, judge advocate.

In case of the absence of any of the members named in this order, the Court will nevertheless proceed with the business assigned it: *provided*, the number be not less than the minimum prescribed by law.

Upon adjournment of the Court *sine die* Lieutenant-Colonel Green, Lieutenants Murphy, Calhoun, Hasbrouck and Cabell, will rejoin their proper stations.

The travel required in the execution of this order is necessary for the public service.

By command of Brigadier-General GIBBON:

H. CLAY WOOD,
Assistant Adjutant General.

OFFICIAL:

John P. Wisser

Aid-de-Camp.

HEADQUARTERS DEPARTMENT OF THE COLUMBIA,
Vancouver Barracks, W. T., April 16, 1886.

SPECIAL ORDERS }
No. 62. }

4.—A general court-martial is appointed to meet at Fort Cœur d'Alêne on Tuesday, the 20th instant, at 10 o'clock, a. m., or as soon thereafter as practicable, for the trial of such prisoners as may be properly brought before it.

DETAIL FOR THE COURT:
1. Colonel *Frank Wheaton*, 2d Infantry.
2. Captain *Curtis E. Munn*, Assistant Surgeon.
3. Captain *Abner Haines, jr.*, 2d Infantry.
4. Captain *James Ulio*, 2d Infantry.
5. 1st Lieutenant *Horace B. Sarson*, 2d Infantry.
6. 1st Lieutenant *William J. Turner*, 2d Infantry.
7. 1st Lieutenant *Richard T. Yeatman*, 14th Infantry.
8. 1st Lieutenant *William R. Abercrombie*, 2d Infantry.
9. 2d Lieutenant *Frederic S. Calhoun*, 14th Infantry.
10. 2d Lieutenant *Stephen J. Mulhall*, 14th Infantry.
11. 2d Lieutenant *Abner Pickering*, 2d Infantry.
12. 2d Lieutenant *Frederick T. Van Liew*, 2d Infantry.
13. 2d Lieutenant *Charles D. Towsley*, 2d Infantry.

1st Lieutenant *John Kinzie*, 2d Infantry, judge advocate.

In case of the absence of any of the members named in this order, the court will nevertheless proceed with the business assigned it: *provided*, the number be not less than the minimum prescribed by law.

Upon adjournment of the court *sine die* Lieutenants Yeatman, Calhoun and Mulhall will rejoin their proper station,—Vancouver Barracks.

The travel required in the execution of this order is necessary for the public service.

By command of Brigadier-General GIBBON:

H. CLAY WOOD,
Assistant Adjutant General.

OFFICIAL:

John P. Wisser

Aid-de-Camp.

Left Vancouver Bks with Co F. in Expd'tion
to Seattle W.T. per orders No. 29 Vancouver
Barracks W.T. Febry 8th 86. On Febry 9th
1886. Arrived at Seattle Feby 10th 1886
Return to Vanc. Bks Febry 25th 86

Knights of labor riots against Chinese
was the cause of these movements of troops.

[Here Calhoun has pasted two typed orders: Special Orders No. 43 and Special Orders No. 62, both of which are orders for the assembly of General Courts Martial. Calhoun is the 9th listed officer for each court. S.O. 43 is issued from Vancouver Barracks on March 15, 1886; S.O. 62 is issued from Vancouver Barracks on April 16, 1886. Transcriptions of each are on the two following pages.]

HEADQUARTERS DEPARTMENT OF THE COLUMBIA,
Vancouver' Barracks, W. T., March 15, 1886.
Special Orders }
   NO. 43

7.—A General Court-Martial is appointed to meet at Fort Walla Walla on Monday, the 22d instant, at 10 o'clock, a.m., or as soon thereafter as practicable, for the trial of such prisoners as may be properly brought before it.
   DETAIL FOR THE COURT:
1. Lieutenant-Colonel *John Green*, 2d Cavalry.
2. Major *Thomas McGregor*, 2d Cavalry.
3. Captain *James N. Wheelan*, 2d Cavalry.
4. Captain *Eli L. Huggins*, 2d Cavalry.
5. Captain *Samuel T. Hamilton*, 2d Cavalry.
6. 1st Lieutenant *John Murphy*, 14th Infantry
7. 1st Lieutenant *Charles B. Schofield*, 2d Cavalry.
8. 1st Lieutenant *Benjamin Munday*, Assistant Surgeon.
9. 2d Lieutenant *Frederic S. Calhhoun*, 14th Infantry.
10. 2d Lieutenant *Thomas J. Lewis*, 2d Cavalry.
11. 2d Lieutenant *Frederick D. Holton*, 2d Cavalry.
12. 2d Lieutenant *Alfred Hasbrouck, jr.*, 14th Infantry
13. 2d Lieutenant *Henry C. Cabell, jr.*, 14th Infantry
    1st Lieutenant *Frederick W. Kingsbury*, 2d Cavalry, judge advocate.

In case of the absence of any of the members named in this order, the Court will nevertheless proceed with the business assigned it: *provided*, the number be not less than the minimum prescribed by law.

Upon adjournment of the Court, *sine die* Lieutenant-Colonel *Green*, Lieutenants *Murphy, Calhoun, Hasbrouck* and *Cabell* will rejoin their proper stations.

   By Command of Brigadier-General Gibbon:
            H. CLAY WOOD,
         Assistant Adjutant General
OFFICIAL:
      [signed 'John P. Misser"]
                  Aide-de-Camp

HEADQUARTERS DEPARTMENT OF THE COLUMBIA,
Vancouver' Barracks, W. T., APRIL 16, 1886.
Special Orders }
  NO. 62
  4.—A General Court-Martial is appointed to meet at Fort Coeur d'Alene on Tuesday, the 20th instant, at 10 o'clock, a.m., or as soon thereafter as practicable, for the trial of such prisoners as may be properly brought before it.
DETAIL FOR THE COURT:
1. Colonel *Frank Wheaton*, 2d Infantry.
2. Captain *Curtis E. Munn*, Assistant Surgeon.
3. Captain *Abner Haines, jr.*, 2d Infantry.
4. Captain *James Ulio*, 2d Infantry.
5. 1st Lieutenant *Horace B. Sarson*, 2d Inantry.
6. 1st Lieutenant *William J. Turner*, 2d Infantry
7. 1st Lieutenant *Richard T. Yeatman*, 14th Infantry.
8. 1st Lieutenant *William R. Abecrorombie* , 2d Infantry.
9. 2d Lieutenant *Frederic S. Calhhoun*, 14th Infantry.
10. 2d Lieutenant *Stephen J. Mulhall*, 14th Infantry.
11. 2d Lieutenant *Abner Pickering*, 2d Infantry.
12. 2d Lieutenant *Frederick T Van Liew*, 2d Infantry
13. 2d Lieutenant *Charles D. Towsley*, 2d Infantry
  1st Lieutenant *John Kinzie,y*, 2d Infantry, judge advocate.
  In case of the absence of any of the members named in this order, the Court will nevertheless proceed with the business assigned it: *provided*, the number be not less than the minimum prescribed by law.
  Upon adjournment of the Court, *sine die* Lieutenants *Yeatman, Calhoun,* and *Mulhalll*, will rejoin their proper station--Vancouver Barracks.
  the travel required in the execution of this order is necessary for the public service.
  By Command of Brigadier-General Gibbon:
    H. CLAY WOOD,
    Assistant Adjutant General
OFFICIAL:
    [signed 'John P. Misser"]
  Aide-de-Camp

HEADQUARTERS FOURTEENTH INFANTRY,
Vancouver Barracks, W. T. September 7, 1886.

ORDERS }
No. 81. }

It becomes my sad duty to announce to the regiment the death of its Colonel, Brevet Brigadier-General LEWIS CASS HUNT, which occurred at Fort Union, N. M., yesterday, the 6th instant.

General HUNT graduated from the Military Academy in 1847, and soon entered upon active service in the war with Mexico as a 2d Lieutenant of the 4th Infantry. It was during this war that he contracted a disease from which he has since been a sufferer, and which finally terminated fatally.

He participated in the War of the Rebellion, being appointed Colonel of the 92d New York Volunteers in May, 1862; engaged in the siege of Yorktown, the Peninsula Campaign, and, at the battle of Fair Oaks, was severely wounded. He was appointed a Brigadier-General of Volunteers in 1862; took part in the operations about Suffolk, the several expeditions to the Blackwater, N. C., and participated in the battle of Kinston, N. C., and for gallant and meritorious services in these various engagements, he received brevets up to the grade of Brigadier-General in the Regular Army.

General HUNT was promoted Colonel of the 14th Infantry in 1881, and during the comparatively short time he has been identified with it, he has, by his kindly manner and amiable qualities, endeared himself to the regiment.

On the day succeeding the receipt of this order the flag will be displayed at half-staff, the colors draped, and the officers of the regiment will wear the usual badge of mourning for thirty days.

Lieutenant-Colonel, Commanding.

HEADQUARTERS FOURTEENTH INFANTRY,
Vancouver Barracks W.T., September 1, 1886.

It becomes my sad duty to announce to the regiment the death of its Colonel, Brevet Brigadier-General Lewis Cass Hunt, which occurred at Fort Union, N. *M.* yesterday, the 6th instant.

General Hunt graduated from the Military Academy in 1847, and soon entered upon active service in the war with Mexico as II 2d Lieutenant of the 4th Infantry. It was during this war that he contracted a disease from which he has since been a sufferer, and which finally terminated fatally. He participated in the War of the Rebellion, being appointed Colonel of the 92d New York Volunteers in May, 1862; engaged in the siege of Yorktown, the Peninsula Campaign, and, at the battle of Fair Oaks, was severely wounded. He was appointed a Brigadier-General of Volunteers in 1862; took part in the operations about Suffolk, the several expeditions to the Blackwater, N.C.,and participated in the battle of Kinston,N. C., and (or gallant and meritorious services in these various engagements, he received brevets up to the grade of Brigadier-General in the Regular Army.

General HUNT was promoted Colonel of the 14th Infantry in 1881, and during the comparatively short time he has been identified with it, he has, by his kindly manner and amiable qualities, endeared himself to the regiment..

On the day succeeding the receipt of this order the flag will be displayed at half-staff, the colors draped, and the officers of the regiment will wear the usual badge of mourning for thirty days.

Lieutenant-Colonel, Commanding.

[Brevet Brigadier-General Lewis Cass Hunt was soon replaced by Colonel Anderson, who commanded the unit into the early 20th Century.]

HEADQUARTERS DEPARTMENT OF THE COLUMBIA,
Vancouver Barracks, W. T., October 14, 1886.

SPECIAL ORDERS
No. 181.

2—Lieutenant Frederic S. Calhoun, ---- Infantry, will proceed to Fort Townsend and report to the commanding officer for Garrison Court-Martial duty, upon completion of which he will rejoin his proper station,—Vancouver Barracks.

The travel required in the execution of this order is necessary for the public service.

By command of Brigadier-General GIBBON:
E. J. McCLERNAND,
Acting Assistant Adjutant General.

OFFICIAL:

Acting Assistant Adjutant General.

Orders:
No. 126.

Vancouver Barracks W.T.
September 8th 1886.

Extract.

1 Leave of absence for fifteen (15) days (for hunting & fishing) is hereby granted 2d Lieut. F. S. Calhoun, 14th Infantry.

By order of Lieut. Colonel DeRussy
(sgd.) Jas. A. Buchanan.

Official:
R. T. Yeatman
1st Lieut. & Adjutant 14th Infantry,
for Post Adjutant.

HEADQUARTERS DEPARTMENT OF THE COLUMBIA,
Vancouver' Barracks, W. T., October 14, 1886.
Special Orders }
   NO. 181
   2 - <u>Lieutenant Frederc S. Calhoun</u>, 14th Infantry, will proceed to Fort Townsend and report to the commanding officer for Garrison Court-Martial duty, upon completion of which he will rejoin his proper station, - Vancouver Barracks.
By Command of Brigadier-General Gibbon:
              E. J. McCLERNAND,
          Acting Assistant Adjutant General
OFFICIAL:
        [signed 'E. J. McClernand"]
       Acting Assistant Adjutant General

[New Entry, pasted in and handwritten]

Order   Vancouver Barracks W.T.
No 126   September 8th 1880
                           Extract
I. Leave of absence for fifteen (15) days (for hunting and fishing) is hereby granted 2d Lieut. F. S. Calhoun, 14th Infantry.

                     By order of Lieut. Colonel DeRussy
                         (sgd) Jas. A. Buchanan

Official: R.T. [illegible], 1st Lieut & Adjutant 14th Infantry, Post Adjutant.

HEADQUARTERS 14TH INFANTRY.
Vancouver Barracks, W. T., December 6th, 1886.

ORDERS
No. 103.

1.—After more than twenty-three years of faithful service, *1st Lieutenant Albert Austin*, of our Regiment, has laid down the burden of life. He received painful wounds in War, and worked hard and efficiently in Peace. In all his army life, he is not known to have uttered an unkind word or to have done an unkind deed. We are called upon to mourn his loss as comrade, officer and friend. Brave, charitable, and kind in life, he now sleeps the sleep of the just.

The usual mourning will be worn by the officers of the regiment, for thirty days, and the regimental colors will be appropriately draped.

By order of Colonel ANDERSON:

JAS. A. BUCHANAN,
1st Lieut. and Adjt. 14th Infantry,
~~Post Adjutant~~.

OFFICIAL:

[signature: Jas. A. Buchanan]
1st Lieut. and Adjt. 14th Infantry,
~~Post Adjutant~~.

HEADQUARTERS 14TH INFANTRY.
Vancouver Barracks, W. T., December 6th, 1886.

Orders}
No. 103.}

1.-After more than twenty-three years of faithful service, *1st Lieutenant Albert Austin,* of our Regiment, has; laid down the burden of life. He received painful wounds in War, and worked hard and efficiently in Peace. In all his army life, he is; not known to have uttered an unkind word or to have done an unkind deed. We arc called upon to mourn his loss as comrade, officer and friend. Brave, charitable, and kind in life, he now sleeps the sleep of the just.

The usual mourning will be worn by the officers; of the regiment, for thirty days, and the regimental colors will be appropriately draped.

By order of Colonel Anderson:
JAS. *A.* BUCHANAN,
1st Lieut. and Adjt. 14th Infantry,

Official:

1st Lieut. and Adjt. 14th Infantry,

Orders
N:2

Vancouver Barracks, W.T.
January 5" 1887.

Extract.

I. 2nd Lieut. F. S. Calhoun, 14" Infty, is hereby placed on temporary duty with Co. "D" 14" Infty. He will report at once to the Commanding Officer of that Company for duty.

By order of Colonel Anderson

(sgd) Jas. A. Buchanan
1st Lieut. & Adjt. 14" Infty
Post Adjutant.

Official:
Jas A Buchanan
1st Lieut. & Adjt. 14" Infty.
Post Adjutant.

HEADQUARTERS DEPARTMENT OF THE COLUMBIA,
Vancouver Barracks, W. T., September 11, 1886.

SPECIAL ORDERS
No. 161.

1.—A Board of Officers to consist of—
Captain *Gilbert S. Carpenter*, 14th Infantry,
Captain *George W. Davis*, 14th Infantry,
2d Lieutenant *Frederic S. Calhoun*, 14th Infantry,
will assemble at Vancouver Barracks on Monday, the 13th inst., for the purpose of examining and testing the rifle sight, as modified by the Ordnance Department, and the one modified by Captain *Andrew H. Russell*.

The Board will thoroughly test both sights, and report in full the qualities of each, especially in regard to the overcoming the tendency to "jump" of the hausse slide, when the rifle is fired.

The Board will meet from day to day on the call of the President.

The junior member will record the proceedings.

BY COMMAND OF BRIGADIER-GENERAL GIBBON:
H. CLAY WOOD,
Assistant Adjutant General.

OFFICIAL:
H. Clay Wood.

Assistant Adjutant General.

Orders   Vancouver Barracks, W.T.
No 2   January 5th 1887

Extract

I. 2nd Lieut F.S. Calhoun, 14th Infty is hereby placed on temporary duty with Co. "D." 14th Infty and will report at once to the Commanding Officer of that Company for duty.

By order of Colonel Anderson
(sgd) Jas. A. Buchanan,
1st Lieut. and Adjt 14th Infty,
Post Adjutant

Official:
Jas A. Buchanan,
1st Lieut.
and Adjt. 14th Infty.,
Post Adjutant.

HEADQUARTERS DEPARTMENT OF THE COLUMBIA,
Vancouver Barracks, W. T., September 11, 1886.
Special Orders }
  NO. 161
    1. -A Board *of* Officers to consist *of-*
    Captain *Gilbert S. Carpenter,* 14th Infantry,
    Captain *George W. Davis,* 14th Infantry.
    2d Lieutenant <u>*Frederic S. Calhoun,*</u> 14th Infantry,
will assemble at Vancouver Barracks on Monday, the 14th inst., for the purpose of examining and testing the rifle sight, 88 modified by the Ordnance Department, and the one modified by Captain *Andrew H. Russell.*

The Board will thoroughly test both sights, and report. in full the qualities of each, especially in regard to overcoming the tendency to "jump" of the hausse slide, when the rifle is fired. ·

The Board will meet from day to day on the call of the President.

The junior member will record the proceedings.
By COMMAND OF BRIGADIER-GENERAL GIBBON:
H. CLAY WOOD,
Assistant Adjutant General.
OFFICIAL
Assistant Adjutant General.

HEADQUARTERS DEPARTMENT OF THE COLUMBIA,
Vancouver Barracks, W. T., March 28, 1887.

SPECIAL ORDERS }
No. 50.   }

1.—1st Lieutenant *Frederic S. Calhoun*, 14th Infantry, will proceed to Fort Townsend and report to the commanding officer for garrison court-martial duty, upon completion of which he will rejoin his proper station,—Vancouver Barracks.

The quartermaster's department will furnish the necessary transportation.

By command of Brigadier-General GIBBON:
                        WM. J. VOLKMAR,
                        Assistant Adjutant General.

OFFICIAL:

            Aide-de-Camp. {Left Post Mch 30
                              {Returns April 4"

HEADQUARTERS DEPARTMENT OF THE COLUMBIA,
Vancouver Barracks, W. T., September 24, 1887.

SPECIAL ORDERS }
No. 145.   }

1.—1st Lieutenant *Frederic S. Calhoun*, 14th Infantry, will proceed to Fort Townsend and report to the commanding officer for duty with Company A, 14th Infantry.

The travel required in the execution of this order is necessary for the public service.

By command of Brigadier-General GIBBON:
                        WM. J. VOLKMAR,
                        Assistant Adjutant General.

OFFICIAL:

            Aide-de-Camp.

HEADQUARTERS DEPARTMENT OF THE COLUMBIA,
Vancouver Barracks, W. T., October 15, 1887.

SPECIAL ORDERS }
No. 154.   }

1.—1st Lieutenant *Frederic S. Calhoun*, 14th Infantry, is relieved from duty as member of the General Court-Martial convened at Vancouver Barracks by virtue of paragraph 5, Special Orders No. 125, current series, from these Headquarters, to take effect upon completion of any case which may be before the Court upon receipt of this order.

By command of Major-General HOWARD:
                        WM. J. VOLKMAR,
                        Assistant Adjutant General.

OFFICIAL:

            Aide-de-Camp.

[It is interesting that while Calhoun was so meticulous in posting copies of travel orders and special duty assignment orders, he did not post copies of orders promoting him to First Lieutenant. On page 79 of his log book, Calhoun posted three sets of typewritten orders assigning him to courts martial and temporary duty in Fort Townsend. Transcriptions of these orders are copied on this page and the following two pages.]

HEADQUARTERS DEPARTMENT OF THE COLUMBIA,
Vancouver' Barracks, W. T., September 24, 1887.
Special Orders }
  NO. 145

    1.—1st Lieutenant Frederic S. Calhoun, 14th Infantry will proceed to Fort Townsend and report to the commanding officer for duty with Company A, 14th Infantry.

    The travel required in the execution of this order is necessary for the public service.

    By Command of Brigadier-General Gibbon:

              Wm. J. VOLKMAR,

            Assistant Adjutant General

    OFFICIAL:

              Aide-de-Camp
              [hand written note:
              This order was suspended by Lt Patterson getting a delay.
              Left Vanc. Oct 19th 87]

HEADQUARTERS DEPARTMENT OF THE COLUMBIA,
Vancouver' Barracks, W. T., October 15, 1887
Special Orders }
   NO. 154

   1.—1st Lieutenant Frederic S. Calhoun, 14th Infantry, is relieved from duty as member of the General court-Martial convened at Vancouver Barracks by virtue of Paragraph 5, Special Orders No. 125, current series, from these Headquarters, to take effect upon completion of any case which may be before the Court upon receipt of this order.

  By Command of Brigadier-General Gibbon:
        Wm. J. VOLKMAR,

      Assistant Adjutant General

OFFICIAL:

                Aide-de-Camp

Post of Fort Townsend, W.T.
October 19th 1887

Orders No. 86.

1st Lieut. F. S. Calhoun 14th Infantry, having reported at post under the provision s of S.O. No 145. Dep't of the Columbia C. S., is hereby attached to company "a" 14th Infantry, for duty.

By order of Captain Bainbridge.
(sgd) Wm. B. Reynolds.
2d Lieut. 14th Infantry,
Post Adjutant.

Official.

2d Lieut. 14th Infantry,
Post Adjutant.

> 86
>
> Post of Fort Townsend, W.T.
> October 19th 1887.
>
> Orders No 86.
>
> 1st.Lieut.F.S.Calhoun 14th Infantry, having reported at post under the provisions of S.O.No 145.Dep't of the Columbia C.S.,is hereby attached to Company "A" 14th Infantry, for duty.
>
> By order of Captain Bainbridge.
>
> (sgd) Wm.B.Reynolds.
>
> 2d.Lieut.14thInfantry,
> Post Adjutant.
>
> Official.
>
> *[signature]*
> 2d.Lieut.14thInfantry,
> Post Adjutant.

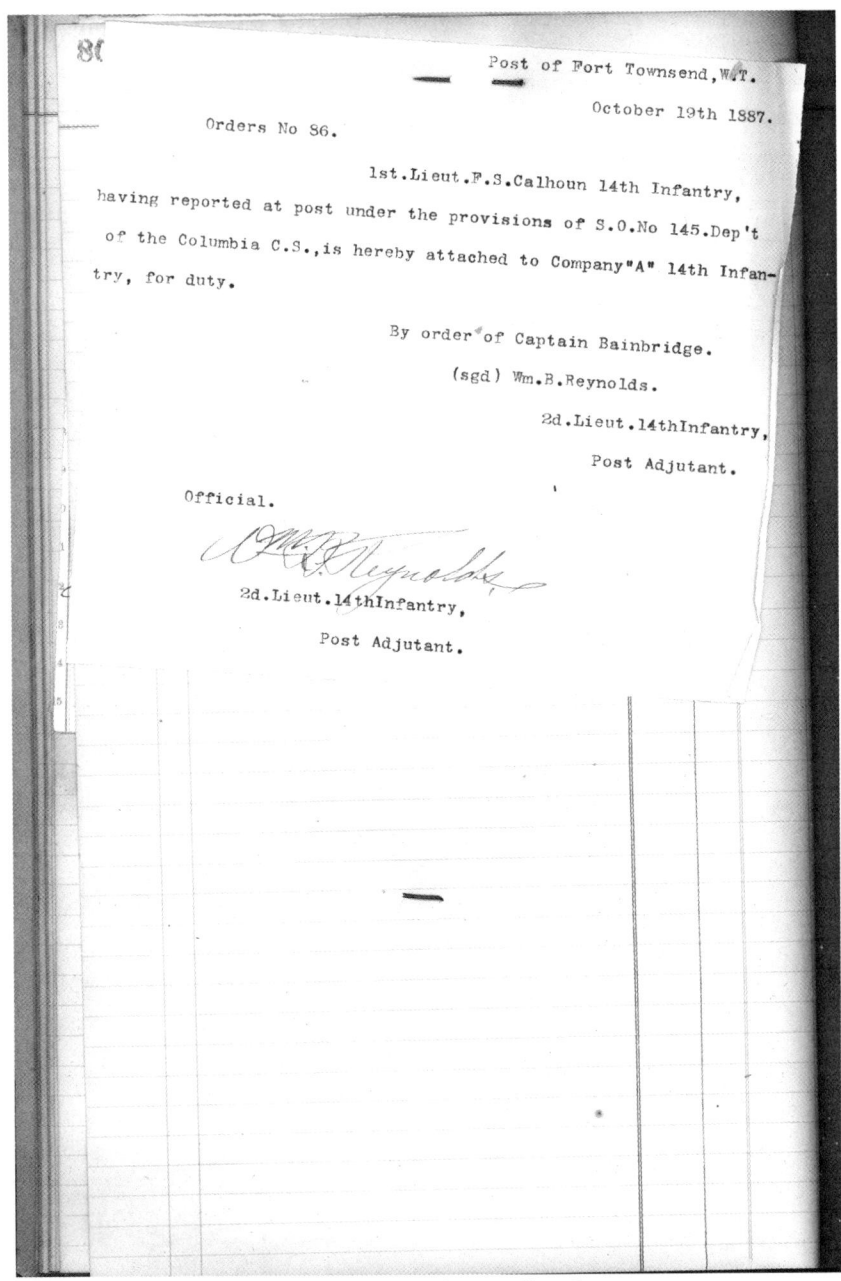

[Here Calhoun has pinned a typed order reading:]

Post of Fort Townsend, W.T.

October 19th 1887.
Orders No 86.

1st.Lieut.F.S.Calhoun 14th Infantry, having reported at post under the provisions of S.O. No 145.Dep't of the Columbia C.S., is hereby attached to Company "A" 14th Infantry, for duty.

By order of Captain Bainbridge
(sgd) Wm.B.Reynolds.
2d.Lieut.14thInfantry,
Post Adjutant.

Official,

2d Lieut. 14th Infantry
Post Adjutant.

[The following seven pages contain a report describing the establishment of a canteen at the Vancouver Barracks. In more isolated posts, it was typical to allow a civilian contractor, generally known as a suttler, open a combination store-tavern within or adjacent to the post. While it granted a monopoly to the contractors to charge whatever they wanted for their goods, at the same time, having a single location for soldiers to recreate at granted a measure of control for the Post Commander. (Habagood and Skaer, 1994)

As time passed, large communities grew up around posts, like the Vancouver Barracks. With outside competition, many of the suttlers went broke. While soldiers could leave post and procure fresh food and drink at lower prices, they were also victimized by many of the bar owners whose businesses the soldiers visited. With so many locations to drink, it was difficult for the senior leadership at the Vancouver Barracks to find their soldiers for some time after payday.

This canteen was a well thought out plan to provide inexpensive food, drink, and a safe recreation site within the post boundaries for soldiers. It also helped enforce discipline. Throughout the month, soldiers were granted credit at the canteen (which was not available at most civilian bars.) If the soldiers drank too much, they were banned from the canteen, and their credit was revoked.

The canteen was run and owned by the various units stationed at the Barracks. Initially it's finances were guaranteed by the officer's on post. However, it very quickly turned a profit. Part of this profit was saved as operating funds, another part of the fund was used for various projects that benefited the soldiers. An example of this was purchasing billiard tables, playing cards, furniture and other recreational items for free use of the soldiers at the canteen.

This model was so successful that it became the genesis of the Army Air Force Exchange Services (AAFES) otherwise known as the Post Exchange or simply PX. The PX system has retail outlets in virtually every U.S. Armed Forces post ascross the world. Today, it still returns a share of its profits to the post and service members it serves.]

Custer's Other Brother-in-Law

VANCOUVER BARRACKS, WASH. T., *August 10, 1886.*
COL. I. D. DE RUSSY,
   Commanding 14th Infantry.

SIR: In accordance with your verbal request, I have the honor to make the following report on the operations of the regimental canteen established at this post last January:

ITS NECESSITY.

Vancouver Barracks has no post trader, nor has there been one here since about 1877, its proximity to the town of Vancouver, with its numerous stores and saloons, rendering the position one of but slight profit. Indeed, the last post trader's store was little more than a bar-room, where the enlisted men spent but little cash, but exhausted all possible credit.

With the post tax, pay of employés, and his own profit to be made, the post trader could hardly compete with dealers in town, who hold out every inducement to the soldiers, and who can afford to sell their goods at a much smaller profit.

The town contains a large number of drinking saloons, running up to the very edge of the military reservation, which have reaped on every pay day a rich harvest for the vile whisky and other liquors sold to the soldiers.

A CANTEEN STARTED.

About the first of January last, under your instructions, preparations were made for inaugurating a canteen at this post, and the necessary supplies for furnishing lunches, coffee, tobacco, cigars, and beer were obtained.

The question at once arose as to the responsibility for the necessary obligations to be incurred, as well as to the ownership of the property to be acquired. It was decided to fix the ownership in the regiment, and that the institution should be known as "The 14th Infantry Regimental Canteen." The reasons for this were, first, that all debts and responsibilities were assumed by the officers of this regiment; second, that the necessary labor was performed voluntarily by the officers of the regiment; and third, that the entire regiment was then at this post, and comprised ten-elevenths of the whole command.

The canteen opened on the sixth of January, and on that day the troops were paid. The articles furnished for sale were cigars, smoking and chewing tobacco, sandwiches made of ham, bologna sausage and cheese, cake, pies of various kinds, coffee, milk, ginger beer, mineral water, and beer.

1 Lt. Calhoun, 14 Inf.

The lunch-counter was put up in the so-called "amusement room," which already contained two rented billiard tables and the periodicals furnished by the Quartermaster's Department.

Small tables were scattered about the room, and cards, checkers, backgammon, chessmen, and other games were furnished in ample quantities.

PRICES.

The absolute absence of any currency less than a five-cent piece rendered it impossible to fix a scale of prices upon any other basis than five cents, or a multiple of that sum, and as a result nearly every article mentioned was sold at five cents. Beer was sold at five cents a glass, or fifteen cents a bottle; cigars, at five cents apiece, but a superior cigar was offered at two for fifteen cents. Plug tobacco was cut into small pieces, but a whole plug costing forty cents was sold for forty-five; smoking tobacco costing forty-eight was sold for fifty, and that costing sixty-five and seventy cents, in half-pound packages, at seventy and eighty cents per pound, respectively. Five cents does not cover the cost of a sandwich, but as a cup of coffee or a glass of beer is almost invariably sold with it, it has been deemed best to continue its sale at that price. Five cents a game is charged for billiards, but all other games are furnished free.

SUCCESS OF THE UNDERTAKING.

As has been stated, the canteen was opened on the January pay day, and its first day's success was in itself sufficient to place the venture on a safe foundation.

In a few hours after opening, the supplies, which were thought to be amply sufficient, were exhausted, and additional supplies had to be obtained.

The sales the first day were $71.90; the second day, $99.80; the third, $70.20; the fourth, $53; and the fifth, $59.30. The next day was Sunday, on which day beer and billiards were interdicted, and the receipts fell to $15.85, but they rose to $36.05 the day following.

The entire receipts for the month were $756.30, averaging $29.08 per day, including Sundays, when the sales are necessarily small. The receipts for the next month, February, were $445.45, an average of $15.90 per day, a considerable falling off from the previous month, but surprisingly large for the second month after pay day. The next month, March, in which another pay day occurred, the sales were $1,060.85, an average daily receipt of $34.22. During the month of April the cash sales fell to $438.55, an average of $14.61 per day; but as a credit system was established during this month, the cash receipts do not represent the true amount of sales. Credit was given during the month for nearly $300, but as only cash receipts are considered, this amount is taken up in the following month, when paid for.

The receipts for May were $1,128.35, an average of $36.39 per day.

In June the cash receipts were $501.55, averaging $16.71 per day, but the credit sales were over $200.

The July receipts were $1,168.53, a daily average of $37.69.

It thus appears that for the first seven months of its operations the total receipts at the canteen have been $5,499.58, a daily average of $26.08.

SIZE OF THE GARRISON.

When the canteen was inaugurated the garrison consisted of the 14th Infantry entire and Light Battery E, 1st Artillery. Shortly after eight companies were ordered away on detached service, four of them, however, soon returning.

The other four were absent about one month, when two of these returned, the remaining two having continued absent up to the present time, and one additional company was permanently detached on July 1st. It will thus be observed that during the existence of the canteen there has never been present the entire strength of the garrison.

CHECKS AND CREDIT SYSTEM.

To promote amongst the enlisted men a more prudent use of their money, checks in denominations of 5 and 10 cents were sold at the rate of 60 cents' worth of checks for 50 cents in cash, and $1.25 worth of checks for $1 in cash. This has not been as successful as was anticipated, the cash sales of checks, notwithstanding the increase, not averaging more than about $20 a month.

The impecuniosity of the men soon after pay day, and their frequent requests for credit, suggested the necessity of some system of credit, and consequently, in April, the following plan was adopted: Commencing with the second month after pay day, that is with the first day of the month succeeding the one in which they were paid, credit was given to the amount of $3.50 for non-commissioned officers and $2.50 for privates. The following blanks were issued to company commanders:

VANCOUVER BARRACKS, W. T.,

——— ———, 1886.

Received from Lieut. J. A. Sladen, 14th Infantry, treasurer 14th Infantry Canteen, ——— dollars' worth of checks, to be paid for on next pay day.

——— ———,

———, Co. ———, 14th Infantry.

Countersigned:

——— ———,

1st Sergt., Co. ———, 14th Infty.

In a letter to the company commanders it was suggested that credit be allowed to such men as might be designated by them through their 1st sergeants, and thus be made an additional means of maintaining discipline, by

4

giving credit to good men and refusing it to the incorrigible. The applicant was to obtain the blank, duly filled and signed for the amount allowed, from the 1st sergeant, countersigned by the latter, and on presentation to the treasurer of the canteen should receive the designated amount in checks.

All the company commanders present, except three, acquiesced in this arrangement, and the plan has worked perfectly thus far. It was expressly stipulated that no pecuniary liability should be incurred by company commanders by reason of giving these orders, but that all losses should fall upon the canteen.

Credit has also been given to the men of the three excepted companies above mentioned where they were personally known to the treasurer or the steward, or when they came properly vouched for.

Of five hundred dollars' worth of credit given in the first seven months the actual loss thus far has been—

By desertions (3)	$5 50
By discharge (1)	3 00
Total	8 50
Two men have neglected to pay	4 50
And are under charges; making the loss thus far	13 00

As there has been no direct means taken to compel payment, this, under the circumstances, is a small percentage of loss.

THE EXPENDITURES.

The total receipts and expenditures for the seven months ending July 31st are as follows:

Total receipts	$5,499 58
Paid services and running expenses	$325 90
Billiard tables, furniture, fixtures, and repairs	755 74
Supplies for sales	4,414 50
Balance	3 44
Total	5,499 58

PROPERTY.

The canteen now owns, free from debt, two new Brunswick-Balke-Collender billiard tables of the best quality, with racks, cues, pool outfits, and all fixtures complete; lamps for the same; five saloon tables, refrigerator, gas-stove, and coffee urn; beer glasses, coffee cups and saucers, goblets, silver-plated knives, forks, and spoons; rubber spittoons, mats, and matting; an iron safe, cigar-lighter, a large assortment of games and cards, and all the accessories of a well-appointed lunch and amusement room.

5

IMPROVEMENTS.

Since the canteen has been in operation a lunch-room has been added to the building and the lunch-counter removed from the large room used as a billiard and reading room. This new room has been fitted up conveniently with counters, shelving, and cupboards, partially at the expense of the canteen, and the whole interior painted, the canteen paying for the material and the labor being performed by soldiers.

Articles for sale have been added to the list from time to time, as the wants and necessities of the men became known. Writing facilities were introduced into the reading room, and two stamped envelopes and two sheets of paper are sold for five cents, thus furnishing the means and facilities for writing letters. Cleaning materials for both clothing and accouterments is sold at a small fraction above cost.

Several brands and styles of chewing and smoking tobacco, cigars, cigarettes, and pipes, to suit the various tastes, are kept on hand. Writing paper, envelopes, pens, pencils, blacking brushes, polishing powder, heel ball, button brushes, and matches are among the articles offered for sale.

Comfortable tables, round and rectangular, such as are used in saloons, with a shelf underneath for glasses, have been bought from the funds of the canteen.

MANAGEMENT.

The canteen is managed directly by a non-commissioned officer, designated the steward, assisted by a private.

These men are paid, respectively, 50 and 35 cents a day from the funds of the canteen.

They are under the direct supervision of a commissioned officer, called the treasurer, who receives the funds and makes all the purchases.

Every two months the accounts and management is subjected to two inspections, one by the regimental council, which makes a thorough examination of the accounts of the treasurer, as well as of the business of the canteen generally; the other by a committee of non-commissioned officers of the regiment, who inspect the practical workings of it, and make recommendation of such changes as they think will improve the service and material. This committee has already suggested several changes for the better, and they have been carried out.

All supplies are bought by the treasurer and turned over to the steward. The latter renders a daily report to the treasurer, showing, under such headings as tobacco, cigars, beer, lunches, stationery, &c., the amounts on hand, amounts sold, and amounts received each day, and transfers to him all funds received. These reports are presented to the treasurer every morning, when orders for new supplies or the supplies themselves are given to the steward.

On the last day of the month a consolidated report for the month is made out by the steward, showing the entire transactions for the month. He also makes a monthly property return, showing all property on hand, received during the month, and "dropped" as worn out, broken, lost, &c., during the same period.

The steward keeps a blotter, with columns for the different articles, and as sales are made, either for cash or checks, they are at once entered. If, as is likely to occur, the cash at the close of the day does not exactly tally with the blotter, a column for "cash unaccounted for" is provided on the daily report.

Of course, much must necessarily depend upon the honesty of the employés; but with the system indicated accurately kept, no considerable loss could long exist without discovery.

PRACTICAL WORKINGS.

The canteen is no longer an experiment. After seven months' operations it now has a well-furnished and comfortable lunch and amusement room, with about seven hundred dollars' worth of property, a fair supply of goods, a small credit, and no debt, and the enlisted men have been furnished with the best of articles at the lowest possible rate. The main object has been to furnish the best at the least cost. As an illustration of this, the coffee is made of two parts Java to one part Mocha, and fresh coffee is made twice or three times a day, as the demand for it occurs.

The desire is to make it a club for the enlisted men, solely for their direct comfort and benefit, and that they shall be made to feel that they have an ownership in its property and a voice in its management.

The best indication of its necessity is shown in the favor it has met with. Its abolition now would be keenly felt.

The presence of a sergeant as steward has a wholesome effect in keeping order and in controlling the enlisted men.

No excess in drinking is allowed. If a man displays a disposition to drink to excess, further sale is refused him.

Noisy and boisterous conduct is forbidden and restrained. Not one single case of drunkenness has been reported from the canteen thus far, though one or two instances have occurred where the canteen has been credited with it in order to shield the wrong-doer, but in each instance investigation has shown the charge to be false. Indeed, from statements made me by several officers, I am inclined to believe that the number of guard-house cases of drunkenness usually following pay day have decreased since the establishment of the canteen.

REDUCTION OF PRICES RECOMMENDED.

The last regimental council, in its report on the cantine, recommended that a reserve fund be set aside and put in some safe shape to provide for transporting the property in case of a change of station, or for any unforeseen emergency that may arise, and this recommendation was approved by you. When this sum of from three to five hundred dollars has been set aside, it is then proposed to still further reduce the scale of prices, so that only sufficient shall be realized to pay the actual running expenses.

An insurance against fire has been negotiated for the sum of six hundred dollars.

Very respectfully, your obedient servant,

J. A. SLADEN,
1st Lieut., 14th Infantry,
Treasurer Regimental Cantine.

[Indorsement.]

VANCOUVER BARRACKS, WASH. T., *September 2, 1886.*

Respectfully forwarded to the Adjutant General U. S. Army, Washington, D. C. (through department headquarters.)

This canteen was started because of the exorbitant prices charged, and want of proper places of amusement for enlisted men.

In it amusements of various kinds are furnished, and of a nature that men can find diversion without expense. This keeps men in the garrison and prevents much of the drunkenness that always results from their lounging about bar-rooms. Much less of this now exists than formerly.

About a year ago it became known that, although every facility was given men to purchase from the commissary, many were unable to do so a short time after pay day. The want of funds compelled them to go where credit could be obtained even though the price charged was very great. This is provided for by credit during non-pay-day month.

The working of the canteen has been examined into by both officers and men from other posts, and, as far as heard from, has met with their unqualified approval.

In every respect it appears to me that good has come, with no resulting harm. It meets with my heartiest commendation, and I only hope that the success it has met with heretofore may continue.

I. D. DE RUSSY,
Lieut. Col. 14th Infantry, Commanding Post.

[Ledger started but never used.]

[Ledger started, one entry for some wood, however there is no date.]

Wood Received

On Contract of W. E. Foster
To Amount of Wood Received   Cds Feet Inches

Amt Carried Foward

Custer's Other Brother-in-Law

"Dt [Date]   F.S. Calhoun in a/c Post Fund   Cr [Credit]
              Camp Robinson Neb.
[Followed by table for August and September accounts]

Date			$	¢	Date		$	¢
August	12	Rec'd from St. Lloyd	230	75	Aug 31	Expense Bakery		
"	31	Sales Bread during Month	42	10	"	Pd. 1 Ch[f] Baker 19 days @ 35¢	6	65
					"	1 Ass't Baker 8 days @ 20¢	1	60
					"	A.C.S. 2 Candles 28¢		
		Account Audited by			"	" 1½# Oil 85¢	1	13
		Post Council			"	Pd. Post Librarian 19 days 35¢	6	65
					"	Subscription to Periodicals		
						Appropriated by Post		
						Council June 8" 1877	43	55
						Balance on hand to new a/c	213	27
			272	85			272	85

September 1877

Date			$	¢	Date		$	¢
Sept	1	On Hand from August a/c	213	27	7th	By Appro' 6 3d Cavalry	11	87
	1	Sales Bread	2	50	" " "	9th Infty	1	83
	2	" "	3	50	" " "	14th "	2	97
	3	" "	3	00				
	4	" "		95				
	5	" "	1	50				
	6	" "	3	50				
	7	" "	3	50				
	8	" "	2	50				
	9	" "	1	90				
	10	" "	2	00				
	11	" "	2	50				
	12	" "	2	50				
	13	" "	2	70				
	14	" "	2	20				
	15	" "	3	25				
	16	" "	3	00				
	17	" "	2	40				
	18	" "	2	50				
	19	" "	1	25				
	20	" "	2	00				
	21	" "	1	00				
	22	" "		60				
	23	" "		60		Carried ford	16	36
		Carried to next Page	263	32				

F. J. Calhoun in a/c Post Fund. — 249

Date	Description	Amount	Date	Description	Amount
	Brot forward	363 37		Brot forward	16 36
Sept 24	To Sales Bread	3 00	Sept 30	A.C.S. Bill	3 43
25	" " "	1 50	"	Paddock & Co	60
26	No Sales		"	1 Chf Baker 30 dys @ 35	10 50
27	To Sales Bread	1 50	"	1 asst " 30 " @ 20	6 00
28	" " "	45	"	1 Librarian 30 " @ 35	10 50
29	" " "	1 35		Bal on hand	
30	" " "	1 00		to New a/c	224 73
		272 12			272 12
Oct 1	On hand from Sept %	224 73	Oct 28	1 asst Baker	
1	Sales of Bread	2 70		23 days @ 20 ¢	4 60
2	" "	1 60		1 Chf Baker 23 dys 35	8 05
3	" "	2 00		1 Post Librarian 23 dys 35	8 05
4	" "	1 25		A.C.S. acct	3 45
5	" "	2 00			
6	" "	2 40			
7	" "	1 50			
8	" "	2 00			
9	No Sales			On hand to New a/c	227 98
10	Sales	50			
11	"	1 50			
12	No Sales				
13	" "	50			
14	" "	50			
15	" "	1 50			
16	" "	40			
17	" "	2 00			
18	" "	1 25			
19	" "	1 25			
20	" "	1 00			
21		1 50			
		252 43			252 43

250 Dr		H. Calhoun A/c Post Fund				
Oct	23	To bal brought from page 249				227 98
"	"	" Sales of Bread				1 10
"	24	"	"	"	"	70
"	25	"	"	"	"	1 50
"	26	"	"	"	"	3 00
"	27	"	"	"	"	2 00
"	28	"	"	"	"	2 00
"	29	"	"	"	"	70
"	30	"	"	"	"	2 90
"	31	"	"	"	"	1 35

242 33

\multicolumn{3}{l}{T. J. Calhoun ⁰⁄₀ Post Fund}		Cr	251/2

			$	
Oct 20	Pd to Co H 3" Cav Council appr'		15	88
"	"	" " " E " " "		98
"	"	" " " L " " "	44	80
"	"	" Regtl Fund 3" Cavalry	19	21
"	"	" " " 9" Infty	1	86
"	"	" " " 14" "	4	04
31	Clk Baker 8 days @ 35¢		2	80
"	Post Librarian 8 days @ 35¢		2	80
	Transferred to Lt Johnson 14" Infty Post Treas Camp Robinson Neb		146	96
			247	33

Convened by S.O. No 191. Par. 2 Hdqrs. Dept Mo. Sept 19th"
General Court Martial Cases

Date of trial	Rank	Name	No of Case	Co	Regt
	Private	L. Davis	1.	D.	14" Inf
Nov 8" 83	"	P. Farrell	2.	J.	" "
Nov 13 83	"	Andrew Ramsay	3.	D.	" "
Nov 15 83	"	James Burns	4.	D.	" "
Nov 19 83	"	Daniel Padrick	5.	A.	" "
Nov 22 83	"	Daniel McBride	6.	D.	" "
Dec 7 "83	"	Charles Bruckner	7.	H.	" "
Dec 11 '83	Sergeant	Richard A. Bowers	8.	D.	" "
Jan 5 83	Private	William Buck	9.	A.	" "
Feby 27" 84	"	George C. Cooper	10.	H.	" "
Mar 6" 84	"	James Flynn	11.	J.	7th "
Mar 20 84	"	William Allen	12.	C.	15th "
Mar 21 84	"	Frank Wilson	13.	A.	23rd "
May 21 84	"	James A. Walsh	14.	D.	14" "
" 26 84	"	Charles Stork	15.	D.	14" "

[Table includes date of trial (Nov. 8, 1883-May 26, 1884), rank

Custer's Other Brother-in-Law

name, number of case, company, and regiment.]

[Continuation of table on page 227.]
1883   F.S. Calhoun 2Lt, 14 Iftry Judge Adjutant
Meir   at Cantonment on Uncompahgre Colo."

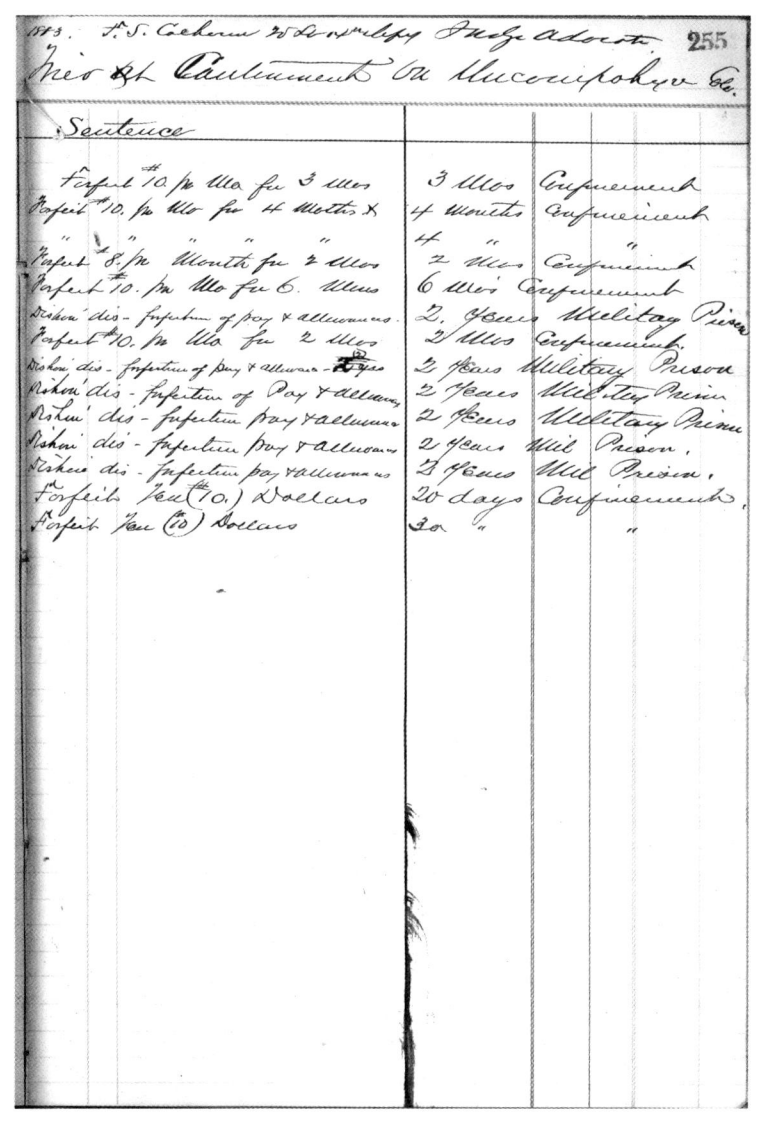

[Table includes sentences of tried persons on page.]

[Here Calhoun has attached a piece of paper that seems to list appointees to a court, which includes his own name. Along the bottom edge of this clipping is a line, probably from a newspaper, that states "How Chinamen Obtain Their Spouses."]

Custer's Other Brother-in-Law

[Here Calhoun has attached a piece of paper that states:]
Proceedings of a Board of Survey convened at Fort Douglas U.T. By S.O. No 1. Hd.Qrs Fort Douglas U.T. / Purpose: / President / Recorder

Napoleons Expedition to Russia — 2 Vols — Genl Count Phillip Segur
Terrible Secret
Under Two Flags — Ouida
Chandos — — — "
Seven Weeks War — 2 Vols — Hozier
Kinglakes Crimea — 3 Vols
Four Years of Fighting — Coffin
Shermans Memoirs — 2 Vols
Napoleon and His Army
We and our Neighbors — H. B. Stowe
Three thousand Miles Thro' the Rocky Mountains — McClure
Farm Legends }
Farm Ballads   } W. Carleton
Harpers History of the Rebellion — 2 Vols
Eugene (Quier) and His time — Mühlbach
Goethe & Schiller — Mühlbach

'77

Two Destinies — Dickens
Ordnance & Gunnery — Benton
Geology — Dana
Testimony of the Rocks — Hugh Miller
Popular Geology — " "

[Possible reading list?]

Napoleons Expedition to Russia - 2. Vols -Genl [illegible] Phillip Logan
Terrible secret.
[illegible]
[illegible]
Seven Weeks War - 2 Vols. Hozier
King lakes Crimea - 3 Vols
Four Years of Fighting - Coffin -
Shermans Memoirs - 2 Vols -
Napoleon and His Army -
We and our Neighbors - H.B. Stowe.
Three thousand Miles Thru the Rocky Mountains - McClure
Farm Legends, Farm Ballads - W. Carleton
Harpers History of the Rebellion - 2 Vols.
 Eugene ([illegible]) and His Times - Muhlback -
Goethe & Schiller - Muhlback -

    '77
Two Destinies - Dieters[?]
Ordnance & Germany - Benton
Geology - Dana
Testimony of the Rocks - Hugh Miller
Popular Geology - "   "

[Handwritten:] Blauchester

[Calhoun has pasted a clipping that reads:] The first Monday in November is the last day upon which to pay the second installment of county taxes

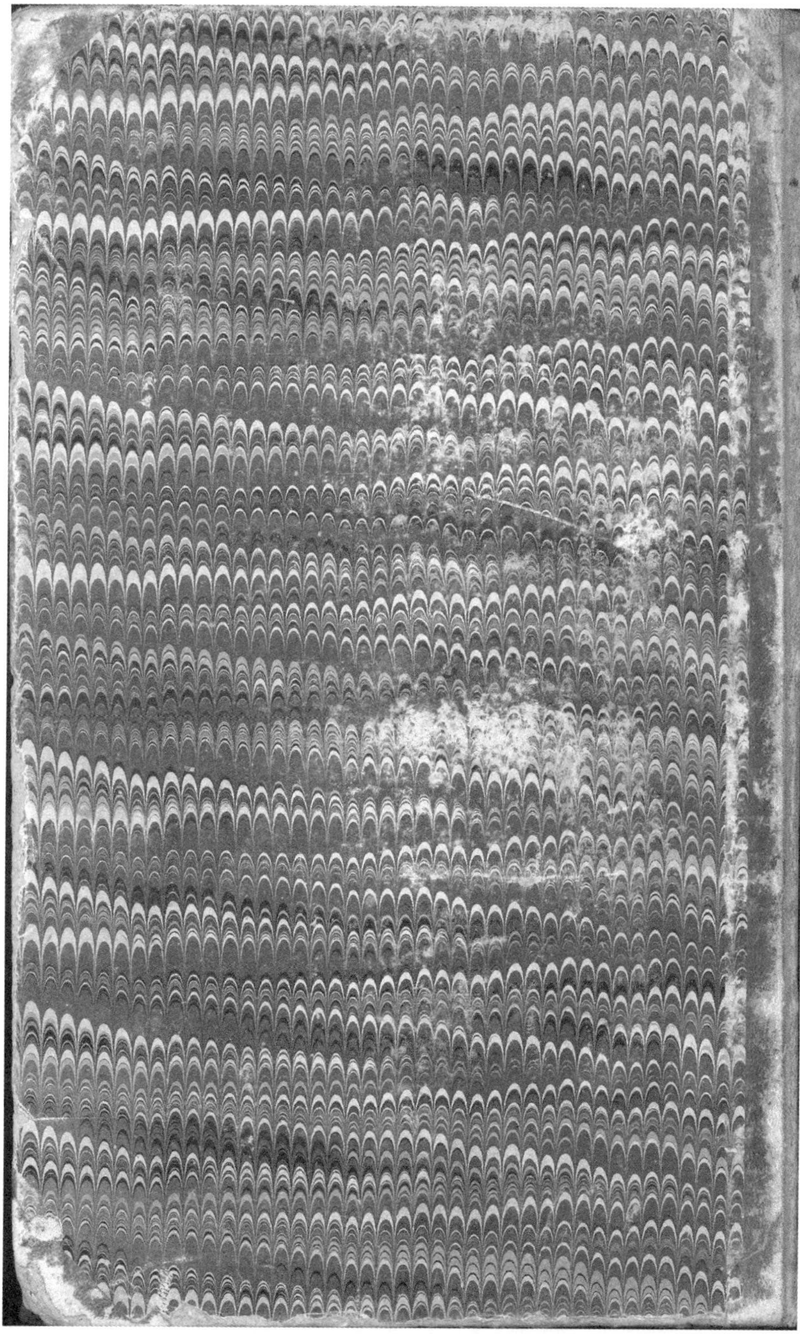

Glossary
of Acronyms and Abbreviations

A.A.AGen	Acting Assistant Adjutant General
AACS	Acting Assistant Commissary Service
A,A,G,/A.A. Gen.	Assistant Adjutant General
A.A.Q.M.	Acting Assistant Quartermaster
A.C.S.	Acting Commissary
Adj't	Adjutant
A.G.O	Adjutant General's Office
AOO	Acting Ordinance Officer
A.Q.M.	Assistant Quarter Master
Asst	Assistant
Bks	Barracks
Bvt	Brevet
Co	Company
Com'd	Command/Commanding/Commander
C.S.	Commissary Service
C.S. Hd. Qrs. G.R.S.	[Unknown]
Decbr	December
Dept	Department
D.T.	Dakota Territory
Expd	Expedition
Ext	Extract
Ex	Extract
Febry	February
Genl	General
G.C.M.	General Court Martial
G.O.	General Order
G.R.S.	[Unknown]
Hdqrs/Hd.Qrs.	Headquarters
Headqrs	Headquarters
Inftry	Infantry
Inst.	*intante mense*, this month
Instant.	*intante mense*, this month

Jany	January
Ky	Kentucky
Lieut	Lieutenant
Maj	Major
Minn	Minnesota
Mo	Missouri
No.	Number
NPPR	Northern Pacific Railroad
NY.N.	New York
O'C	O'clock
O. & G. Equipage	Ordinance & G Equipage [Unclear]
Par	Paragraph
Para	Paragraph
Proximo.	The month following the present.
Q.M.	Quartermaster
Qs Ms	Quartermasters
Rec'd	Received
Rectg Service	Recruiting Service
Reg	Regulation
Rev	Revised
Sgd	Signed
Sine Die	Without delay
S.O.	Special Orders
USRR	U.S. Railroad [Unknown]
U.C.RR.	U.C. Railroad [Unknown]
U.S.A.	United States Army
U.T.	Utah Territory
Van	Vancouver
Vis	Attack
W.T.	Washington Territory
W.T.	Wyoming Territory
Wy	Wyoming

Appendix B
14th Infantry
(Golden Dragons)[5]
Lineage of the 14th Infantry Regiment

<u>Constituted:</u> 3 May 1861 in the Regular Army as the 2d Battalion, 14th Infantry

<u>Organized:</u> 1 July 1861 at Fort Trumbull, Connecticut
Reorganized and redesignated 30 April 1862 as the 1st Battalion, 14th Infantry

Reorganized and redesignated 21 September 1866 as the 14th Infantry

Consolidated 26 July 1869 with the 45th Infantry, Veteran Reserve Corps (constituted 21 September 1866), and consolidated unit designated as the 14th Infantry.

Assigned 27 July 1918 to the 19th Division

Relieved 14 February 1919 from assignment to the 19th Division
Assigned 10 July 1943 to the 71st Light Division (later redesignated as the 71st Infantry Division)

Relieved 1 May 1946 from assignment to the 71st Infantry Division
Inactivated 1 September 1946 in Germany

Activated 1 October 1948 at Camp Carson, Colorado

Assigned 1 August 1951 to the 25th Infantry Division

---

[5] Center of Military History, Carlisle Barracks, VA. Lineage and Honors Information as of 2 June 1997

Relieved 1 February 1957 from assignment to the 25th Infantry Division and reorganized as a parent regiment under the Combat Arms Regimental System

Withdrawn 1 March 1986 from the Combat Arms Regimental System and reorganized under the United States Army Regimental System

Honors of the 14th Infantry Regiment (Golden Dragons)

Campaign Participation Credit

Civil War: Peninsula; Manassas; Antietam; Fredericksburg; Chancellorsville; Gettysburg; Wilderness; Spotsylvania; Cold Harbor; Petersburg; Virginia 1862; Virginia 1863

Indian Wars: Little Big Horn; Bannacks; Arizona 1866; Wyoming 1874

War with Spain: Manila

China Relief Expedition: Yang-tsun; Peking

Philippine Insurrection: Manila; Laguna de Bay; Zapote River; Cavite; Luzon 1899

World War II: Rhineland; Central Europe

Korean War: UN Summer-Fall Offensive; Second Korean Winter; Korea, Summer-Fall 1952; Third Korean Winter; Korea, Summer 1953

Vietnam: Counteroffensive; Counteroffensive, Phase II; Counteroffensive, Phase III; Tet Counteroffensive; Counteroffensive, Phase IV; Counteroffensive, Phase V; Counteroffensive, Phase VI; Tet 69/Counteroffensive; Summer-Fall 1969; Winter-Spring 1970; Sanctuary Counteroffensive;

Counteroffensive, Phase VII; Consolidation I; Consolidation II; Cease-Fire

Decorations
Presidential Unit Citation (Navy) for CHU LAI
Valorous Unit Award for MOGADISHU
Republic of Korea Presidential Unit Citation for MUNSAN-NI

The 14th Infantry Regiment, following the end of the United States Army Regimental System

Two battalions of the 14th Infantry Regiment are currently active. The 1st Battalion, 14th Infantry, is assigned to the 2nd Brigade, 25th Infantry Division, at Schofield Barracks, Hawaii. The 2nd Battalion, 14th Infantry, is assigned to the 2nd Brigade, 10th Mountain Division ,at Fort Drum, New York.

Their military deployments include:

Somalia, 1993

Bosnia, 2002

The Global War on Terrorism, 2003–present

Distinctive Unit Insignia[5]
(See back cover of this book for a full color version)

Description/Blazon
A gold color metal and enamel device 1 1/8 inches (2.86cm) in height overall consisting of a gold imperial Chinese dragon placed against a red conventionalized Spanish castle with the motto "THE RIGHT OF THE LINE" in gold letters on a blue ribbon scroll.

Symbolism
The dragon is the crest of the regiment and the castle is one of the charges on the regimental shield. The motto is the motto of the regiment.

Background
The distinctive unit insignia was originally approved on 6 Nov 1924. It was amended on 11 Jun 1925 to correct the color of the motto letters.

---

[5] Duic and Coat of Arms courtesy the Army Institute of Heraldry.

Coat of Arms 14th Infantry Regiment
(See back cover of this book for a full color version)

Description/Blazon

<u>Shield:</u> Per fess Azure and Argent, two arrows chevronwise point to point counterchanged between in chief a cross patée of the last and in base a spreading palm Vert debruised by a castle.

<u>Crest:</u> On a wreath of the colors an imperial Chinese dragon affronté Or scaled and finned Azure incensed and armed Gules.

<u>Motto:</u> **THE RIGHT OF THE LINE**

Symbolism

<u>Shield:</u>The regiment was organized in 1861 and played a notable part in all the Virginia Campaigns from the Siege of Yorktown in 1862 to October 1864. It was in Sykes' regular division of the 5th Corps of the Army of the Potomac whose badge was a white cross patée. At Gaines Hill and Malvern Hill, the division commander praised the regiment, and the brigade, and commended it at Second Manassas. It performed a most difficult service at Antietam, was in the repulse of the crucial attack of the enemy at Gettysburg and made a most gallant charge at the Wilderness.

In later years, the regiment took part in two Indian Campaigns indicated by the two arrows; and detachments were in two others, but not in sufficient strength to entitle the regiment as a whole to participation. It was at the capture of Manila in the Spanish War indicated by the castle, and in the fighting around the same city in

1899 indicated by the palm, and in the China Relief Expedition as shown by the dragon.

<u>Crest:</u> None

<u>Motto</u>

The motto is the much prized remark made by General Meade directing the station of the regiment in the Review just after the Civil War.

The coat of arms was approved on 10 Dec 1921.

Bibliography

Buecker, Thomas, R. A little-known member of the "Custer clique". *Greasy Grass* Vol. 10, May 1994.

Habagood, Carol A, Colonel and Skaer, Marcia, Lieutenant Colonel, *One Hundred Years of Service: A History of the Army and Air Force Exchange Service, 1895 to 1905*, Headquarters Army and Air Force Exchange Service, Dallas, Texas, 1994

Keen, B. M., *Statement of Military Service of Frederic S. Calhoun*, U.S. War Department, Adjutant General's Office, Jan 4, 1890.

Rodenbough, Theophilus F. Bvt Brigadier General U.S.A. and Haskin, William L. Major, First Artillery, editors. *The Army of the United States Historical Sketches of Staff and Line with Portraits of Generals-in-Chief.* Maynard, Merrill, & Co. New York, 1896

Index

A

Abercrombie, William 196
African American soldiers 62-63
Anderson, Thomas 27, 34, 35, 39, 199, 203-205
Anti-Chinese Riots 20-21, 33, 192-195
Apache Wars 28, 32-33
Argentine flag 48-49
Aspinwall, John 11-12, 27
Austin, Albert 32, 171, 177, Obituary 201-202

B

Bainbridge, A.H. 28-30, 35, 174-175, 208-211
Band, military 17, 28, 29, 62-63, 66-67
Bannack War, 18, 32, 38
Bannock War (see Bannack War)
Belknap, William 11, 12
Benteen, Frederick 5, 9-12, 22
Big Horn and Yellowstone Expedition (see Great Sioux War)
Blunt, Mathew 30, 34
Boards of Survey 7, 58-59, 74-75, 78-83, 98-99, 108-109, 112-123, 138-139, 146-151, 229
Bourke, John 104-105
Bradley, L.P. 118-119, 124-139
Bryant, Montgomery 32, 34, 72-79, 82-83, 86-87, 90-91, 140-141, 152-155, 170-171
Bubb, John 98-99

Buchanan, Jason 35, 200-205
Burke, Daniel 30, 31, 35, 74-79, 82-83, 92-93, 110-113, 150-151
Burke, Paul 80-81
Burrowes, Thomas 130-131, 138-139

C

Cabell, Henry jr. 35, 195-196
Calhoun, Elizabeth 22
Calhoun, James, 5, 6, 9, 17, 37, 54-57
Calhoun Fred, biography 9-26, timeline 37-39, personal notes 54-57, 64-65, 70-71, 77-78, 82-85, 90-91, 94-95, 162-163, 166-169, 192-195, medical retirement 21-22, 39, duty with Fifth Cavalry, 154-155, Commanding Ft Lyon, CO 178-180, rifle sight test 203-204, Regimental Canteen 211-218
Calhoun, Charlotte 9, 15
Camp (Fort) Douglas, UT 13, 14, 18, 19, 33, 37, 38, 41, 47, 70-93, 140-153, 156-175, 229
Camp Robinson 6, 15, 16- 18, 98-101, 106-135, 138-139, 160-161, bakery fund 221-224
Camp Warner, OR 10
Camp Whitewood Cr, DT 96-97
Canteen, Vancouver Barracks 7, 212-218
Cantonment on the Uncompahgre 20, 38, 184-191, 226-227
Carlin, William 12, 13, 50-51
Carling, Elias 146-147
Carpenter, Gilbert 30, 32, 35,

243

Index

78-79, 86-87, 114-115, 176-177, 204-205
Carpenter, William 98-99, 118-123
Chambers, Alexander 14, 31, 96-97
Civil War 9, 10
Clark, William P. 16, 122-123, 160-161
Clements, Bennett 144-145, 148-149, 150-151
Cleutes, Edward 104-105
Coal 78-79, 150-151
Coffee 7, 68-69, 74-75, 214
Collier, William 140-141
Colored soldiers, (see African American soldiers)
Cook, Samuel 104-105
Corinne, UT 14, 76-77, 144-145
Courts Martial 6, 21, 58-63, 72-73, 76-79, 84-89, 92-93, 110-111, 114-115, 130-131, 134-135, 140-145, 150-151, 156-157, 174-175, 194-197, 200-201, 207-209, 226-227, 228, 234
Crazy Horse 6-7, 9, 18, 38
Creamer, John 104-105
Crittenden, Albert 14, 15, 28
Crittenden, John, 14
Crofton, Robert 70-71
Crook, George 6, 14-16, 31, 32, 72-73, 80-81, 86-87, 98-99, 104-107, 132-135, 138-141, 152-153, 156-157, 166-167, 170-171, 174--177, Speech 100-105
Cummings, Joseph 110-111, 132-133
Cusack, Patrick 60-62
Custer, Boston 5, 11, 12, 15
Custer City, DT 96-99

Custer, Elizabeth 11, 19, 22, 23
Custer, George 3, 5, 6, 10, 11-13, 18, 19, 37, 38, 54-57
Custer, Margaret Emma 10, 11, 17, 19, 20, 37
Custer, Tom 11

D

Dandy, George 12, 13, 50-51
Davis, George 29, 30, 35, 204-205
Deadwood, D.T. 15
Desertion 11, 27, 108-109, 112-115, 190-191
Division of the Missouri, troop movements Oct. 1881 180-183
DeRussy, Isaac 34, 35, 200-201
Devine, James 104-105
Donally, Michael 104-105
Dorst, Joseph 112-123, 160-161,
Douglas, Henry 34, 178-179, 186-191
Drum, Richard 192-193
Drum, William 34
Duboise, William 104-105

E

Eastman, Frank 35
Edwards, William 104-105
Evans, Andrew 122-123

F

Featherly, William 104-105
Fitzgerald, Michael J. 68-71
Fitzhenry, Robert 104-105
Flynn, Richard 104-105
Ford, Daniel, 104-105
Fort Abraham Lincoln, NB 11, 12, 19, 22, 37, 50-57
Fort Cameron, UT 31, 88-91,

Index

152-153, 164-165
Fort Coeur d'Alene, IT 21, 196-197
Fort D.A. Russell, 70-71, 162-163
Fort Fetterman 14, 31
Fort Fred Steele 72-73, 156-157
Fort Hall, IT 18, 38, 152-155, 168-169, 174-175
Fort Laramie, WT 30, 110-111
Fort Leavenworth, KS 17, 21, 180-181, 184-185
Fort Lyon, CO 38, 178-179, 180-181, 184-185
Fort Omaha, NB, 13, 17, 66-71, 128-129, 170-171
Fort Townsend, WT 21, 39, 200-201, 206-211
Fort Vancouver (Vancouver Barracks) 2, 3, 20-22, 28, 29, 33, 39, 192-197, 199-203, 206-208, rifle testing 204-205, regimental canteen 212-219
Fort Walla Walla, WT 21, 194-196
Foster, Charles 104-105
Fourteenth Infantry Regiment history 27-36, lineage and honors 236-241
Furey, John 150-151

G

Gentles, William 18, 25
Gibbon, John 22, 194-197, 200-201, 204-208
Glass, Edward 104-105
Goldin, Theodore 22
Goodwin, William P. 35
Governors Island 64-65
Great Sioux War 5, 6, 9, 14-17, 31-32, 37, 94-105

Green, John 195-196
Greene, Oliver 50-53
Gustin, Joseph 35, 80-81, 86-87

H

Haines, Abner, jr. 196-197
Hamilton, J.M. 29, 30, 154-155
Hamilton, Samuel 194-196
Hammond, Charles 130-131
Harold, Horace 104-105
Hasbrouck, A. jr. 35, 194-196
Hasson, Patrick 32, 33, 35, 140-143
Haughey James 56-59, 62-63
Hay 146-149
Henderson, Nevil Mrs. 62-63
Henry, Guy 104-105
Holton, Frederick 194-195
Horton, Samuel 175-176
Huggins, Eli 194-195
Hughes, Robert 52-53
Hunt, Lewis Cass 33, 34, 172-173, obituary 198-199
Huston, Daniel 178-179

I

Ilges, Guido 29, 30, 33
Indian Reservation supplies 112-113, 118-119, 124-125, 130-131

J

Jackson, Henry 22
Johnson, Charles 35, 80-81, 118-119, 130-131
Johnson, Theodore 114-15
Jones, William K. 35
Jordan Creek, Battle of 32

245

Index

K

Kennington, James 31, 72-73
Kimball, William 32, 35, 140-141, 154-155
Kinzie, John 196-197
Kirkwood, Scott 104-105
Krause, David 29, 30, 32, 90-91, 152-153

L

Lascombeskie, John 104-105
Leave 16, 18- 20, 22, 38, 39, 98-99, 106-109, 154-159, 166-167, 184-189, 192-193, 200-201
Lee, John 31
Lewis, Thomas 195-196
Little Big Horn, Battle of 3, 5, 6, 9, 14, 18, 19
Lloyd, Charles 31, 80-81, 86-87, 110-119, 124-135
Loosch, Andrew 104-105
Lovell, Charles 28- 30, 35
Lovell, Robert 90-91, 140-141

M

McAleer, (no first name) 190-191
McCammon, William 30, 35, 46-47, 74-85, 88-93, 142-153, 158-159, 162-173
McCarribe, Samuel 176-177
McConihe, Samuel 32, 35, 86-89, 92-93
McGillycuddy, Valentine 14, 15
McGregor, Thomas 195
McClernand, Edward 200-201
McKernan, Edward 104-105
MacKenzie, R.S. 112-123, 160-161
Madden, William 104-105
Mason, Edwin 56-59, 62-65
Martin, CharlesH. 35
Martin, John 9
Meagher, Thomas 104-105
Merritt, William 162-163
Mills, Anson 15
Misser, John, 194-196
Morris, Louis M. 56-59, 62-65
Morrow, Stanley 16
Mountain Meadow Massacre 31
Moylan, Miles 15
Mulhall, Stephen 35, 194-197
Munday, Benhamin 194-196
Munn, Curtis 194-197
Murphy, John 31, 35, 74-75, 80-83, 86-87, 92-93, 116-117, 120-121, 130-131, 142-149, 194-196
Musicians 62-63, 66-67

N

Neide, Horace 64-65
Newport Barracks, Ky 37, 50-51, 56-67

O

O'Brien, James 104-105
O'Donnell, Patrick 104-105
O'Neil, Joseph 35
Omaha Barracks 128-129, 156-157

P

Patterson G.T. 31, 170-171, 206-207
Pelouze, L.H. 54-55, 60-61
Pickering, Abner 194-197
Platt, Edward 184-185
Post Exchange, (see canteen)
Prisoners 17, 58-59, 62-63, 84-85, 87-89, 92-93, 114-115,

Index

128-131, 140-145, 174-175, 190-191, 194-196
Puget Sound Expedition (see Anti-Chinese Riots)

Q

Quentin, Julius 72-75, 84-87, 174-175, 178-179, 186-191

R

Randall, George M. 120-121, 124-125
Ray, Patrick 10, 11
Recruits 7, 13, 62-73, 88-91, 152-153, 160-161, 164-165, 168-169
Red Cloud Agency 1, 16, 38, 106-107, 112-113, 124-125, 130-131
Reed, Emma 3, 12, 18, 19, 23, 38
Reed, Harry Armstrong 15, 18
Reno, Marcus 5, 9
Reynolds, William Bainbridge 35, 138-139, 174-175, 208-211
Riley, J.F. 58-59, 68-71
Roberts, C.S. 70-71
Rosser, Thomas 10
Ruggles, George 72-73, 80-81
Russell, Andrew 204-205
Ryan, John 9
Rosebud, battle of 14

S

St. Louis Barracks, MO 62-63
Sample, William 35
Sarson, Horace 194-196
Scharmon, Private 88-91
Schneider, Edward 104-105
Schofield, Charles 194-196

Sherman, William T. 158-159, 188-189, 230-231
Simpson, James 108-109
Slim Buttes, Battle of 15, 31
Smith, Algernon, Mrs. 17
Smith, Fran 104-105
Smith, John 30, 33, 46-47, 76-77, 80-85, 88-89, 92-93, 142-143, 146-149, 152-153, 158-169, retirement 172-173
Smith, Michael 12
Snow, E.A. 104-105
Spotted Tail Agency, NE 16, 18, 160-161
Steigers, A.F. 68-69, 70-71
Steiuer, Henry 104-105
Stephenson, J.W. 104-105
Sugar 74-75
Swift, Ebenezer 60-61
Starvation March 6, 15-16, 32

T

Taylor, Frank 31, 32, 35, 140-141, 150-151, 174-177
Terry, Alfred 15, 50-53
Terry, John H. 104-105
Thomas, Henry 140-141
Thompson, John M. 58-61, 66-67
Tobey, Thomas 13, 16, 28, 31, 32, 35, 78-79, 84-85, 106-107, 114-115, 118-119, 124-125, 130-131, 138-139, 140-147, 150-153, 176-177
Tomatoes 98-99, 120-121
Townsend, Edward 54-58, 60-61, 158-159
Towsley, Charles 194-197
Trotter, Frederick 30, 32, 35
Turner, William 194-197

Index

U

Ulio, James 194-196
Ute War 20, 38, 162-163

V

Van Derslice, Joseph 29, 30, 32
Van Liew, Frederick 194-197
Volkmar, William 205-208
Von Luettwitz, Aldophus 104-105
Vroom, Peter 108-109

W

Wallen Henry 29
Warren, Robert 76-83, 174-177
Warrens, Charles 35
Weeds, James 12, 50-51
Wessells, Henry jr. 126-127
Western. Charles 33, 35, 170-171, 176-177
Wheaton, Frank 194-197
Wheelan, James 194-196
White River War (See Ute War)
Williams, Robert 86-87, 98-99, 106-107, 132-135, 138-141, 152-153, 156-157, 166-167, 174-177, 186-187
Winston, Edward. 35
Wood received 220-221
Wood, Henry 194-197, 204-205
Woodward, George 30, 31, 34
Worstershire sauce 98-99
Wikoff, Charles 34, 35

Y

Yates, George 13, 17, 50-51
Yeatman, Richard 31, 35, 80-81, 106-107, 140-141, 174-175, 194-197
Young, George S. 58-63